Aftershock:
The Ancient Cataclysm That Erased Human History

Authored by Brien Foerster
Cover by Hans Messerschmidt
All photos in the book by Brien Foerster or Irene Mendoza

Machu Pic'chu: Inca repair work on top of older megalithic walls

Dedication

A book like this is not the labor of one person, but is the compilation of all of those that came before him. The Inca were one of the greatest cultures of all time, in part due to their social cohesion and mastery of astonishing agricultural systems that stretched from valley floors to the tops of mountains. Although the colonial Spanish tried brutally to extinguish any vestiges of their culture and works, and wrote incorrect histories about them, we know through oral traditions that they did not claim responsibility for the great megalithic works attributed to them by most writers.

The dynastic Egyptians, makers of astonishing temples and other structures of stone, also recorded through their glyph system that their ancestry went much farther back in time than most academics will accept. It is the hope that this book will shed light into a much more complex and intriguing history of our planet than what we have been taught.

To this end, as regards ancient Peru, I wish to thank Dr. Theo Paredes, Willco Apaza, Sr. Juan Navarro, and Sr. Renato Davila Riquelme for their assistance in guiding me through the complex history of that enchanting country. In Bolivia, Antonio Portugal has been a vital teacher, and in Egypt, no one could possibly have had better resources and insights than those graciously and freely given by Stephen Mehler, Mohamed Ibrahim, and the amazing Yousef and Patricia Awyan, as well as the rest of the Awyan family of Giza.

In addition, to my beloved Irene, who has walked with me every step of the way.

Forward by Stephen S. Mehler

In the last decades of the 20th, and the first two decades of the 21st centuries, many researchers have arisen to seriously question accepted academic theories concerning the ancient history and prehistory of humanity, and call for new paradigm shifts about human origins and distinct existence of ancient advanced civilizations.

Brien Foerster has emerged as one of the most active and profound of these researchers. He not only questions academic theories, but actively and vigorously goes to ancient megalithic sites around the world to investigate for himself. Brien has visited these sites in the presence of world-class archaeologists, geologists, engineers of all types, chemists, physicists, and stone masons.

I have had the privilege to be with Brien onsite four times in Egypt, and to Peru, Bolivia and Lebanon, and watch and experience him deeply investigate these areas.

Brien has produced many books and this current one is an excellent work, with his own concise analysis of the literature and onsite observations. It is full of visual evidence of artifacts and sites not seen by most scientists and laypeople. The evidence is in strong support of the theory that a worldwide cataclysm - mentioned in 'great flood' myths of over 250 different cultures - impacted the Earth and humanity ca. 11,500 years ago. This book is a must read for professionals and laypeople alike - well written and presented with ample physical evidence that can no longer be denied or ignored by anyone seriously interested in the true prehistory of humanity.

Stephen S. Mehler, MA

Lafayette, Colorado

May 2016

Contents

1. The Case For Ancient Cataclysm pg.8
2. Peru And Bolivia pg.64
3. Egypt pg.117
4. Lebanon pg.292
5. Closing Thoughts pg.309
6. Bibliography pg.311

1. The Case for Ancient Cataclysm

Most readers have likely heard the story about the 'great flood' from their childhoods, as portrayed in the Bible. The Genesis chapter describes how a vengeful God punishes humankind by literally flooding the surface of the Earth with water, extinguishing most terrestrial life over 40 days and nights of torrential rain. Those who survive are a small, divinely chosen group led by a righteous man named Noah, who not only leads his family but one male and one female of each land-based species onto a ship he built to survive

One version of how Noah's Ark may have looked

the cataclysmic ordeal.

Many believe so strongly in the ancient texts that comprise the Bible, that they accept the great flood as truth. Readers that are more pragmatic may ask, *When did this flood occur?* And, *How is such a meteorological event possible?* Biblical sources are not forthcoming with answers.

The Flood myth is not solely the domain of one specific religious text or belief system, however, and is often a symbolic narrative in which a great flood is sent by a deity, or deities, to destroy civilization in an act of divine retribution. Parallels are often drawn between the flood waters of these myths and the primeval waters found in certain creation myths, as the flood waters are described as a measure for the cleansing of humanity, in preparation for rebirth. Most flood myths also contain a culture hero, who strives to ensure this rebirth. (1)

The flood myth motif is widespread among many cultures, as seen in Mesopotamian flood stories, the Puranas (ancient Hindu texts), in the Greek Deucalion mythology, the lore of the K'iche' and Maya peoples of Central America, as well as the Muisca people of present day Colombia in South America. In fact, there are oral tradition stories pertaining to this concept from antiquity, from cultures of Sumeria, Babylonia, Germany, Ireland, Finland, the Maasai of Africa, Egypt, India, Turkestan, China, Korea, Malaysia, Lao, Australia, Polynesia, and Native people of North America, Mesoamerica and South America… to name just a handful.

'Catastrophism' is the theory that the Earth was affected in the past by sudden, short-lived, violent events, possibly worldwide in scope. (2) The dominant paradigm of modern geology is called 'uniformitarianism' (sometimes described as 'gradualism'), in which slow incremental changes, such as erosion, create the Earth's appearance. This view holds that the present is the key to the past, and that all things continue as they were, from the beginning of the world. Recently a more inclusive and integrated view of geologic events has developed, changing the scientific consensus to accept some catastrophic events in the geologic past.

Mayan text known as the Dresden Codex

This holds that there have been violent and sudden natural catastrophes such as great floods and the rapid formation of major mountain chains, such as the Himalayas and the Andes of South America. Plants and animals living in those parts of the world where such events occurred were often killed off, according to the 19th century French scientist Georges Cuvier. Then, new life forms moved in from other areas. As a result, the fossil record for a region shows abrupt changes in species.

Cuvier's explanation relied solely on scientific evidence rather than biblical interpretation. His motivation was to explain the patterns of extinction and faunal succession that he and others were observing in the fossil record. While he did speculate that the catastrophe responsible for the most recent extinctions in Eurasia might have been the result of the inundation of low lying areas by the sea, from the melting of ice at the end of the last ice age, he did not make any reference to Noah's flood. (3) Nor did he ever make any reference to divine creation as the mechanism by which repopulation occurred following an extinction event. In fact, Cuvier, influenced by the ideas of the European Enlightenment and the intellectual climate of the French revolution, avoided religious or metaphysical speculation in his scientific writings. (4)

Cuvier further believed that the stratigraphic record (layers of deposits) indicated that there had been several of these revolutions, which he viewed as recurring natural events, amid long intervals of stability during the history of life on Earth. This led him to believe the Earth was several million years old. (5) This clearly flew in the face of the predominant religious contention of the time that the planet on which we inhabit was regarded as only being several thousand years old.

Painting of Georges Cuvier

By contrast, in England, where natural theology was very influential during the early 19th century, a group of geologists that included William Buckland and Robert Jameson would interpret Cuvier's work in a very different way. Jameson translated the introduction Cuvier wrote for a collection of his papers on fossil quadrupeds (having four legs), that discussed his ideas on catastrophic extinction, into English and published it under the title *Theory of the Earth*. He added extensive editorial notes to the translation that explicitly linked the latest of Cuvier's revolutions with the biblical flood, and the resulting essay was extremely influential in the English-speaking world. (6)

Buckland spent much of his early career trying to demonstrate the reality of the biblical flood with geological evidence. He frequently cited Cuvier's work even though Cuvier had proposed an inundation of limited geographic extent and extended duration, and Buckland, to be consistent with the biblical account, was advocating a universal flood of short duration. (7) Eventually, Buckland would abandon flood geology in favor of the glaciation theory advocated by Louis Agassiz, who had briefly been one of Cuvier's students. As a result of the influence of Jameson, Buckland, and other natural theology advocates, the 19th century debate over catastrophism took on religious overtones in Britain that were not nearly as prominent elsewhere. (8)

Uniformitarian explanations for the formation of sedimentary rock and an understanding of the immense stretch of geological time, or as the concept came to be known 'deep time,' were found in the writing of James Hutton, who is sometimes known as the father of geology, in the late 18th century. The geologist Charles Lyell built upon Hutton's ideas during the first half of 19th century and amassed observations in support of the uniformitarian idea that the Earth's features had been shaped by same geological processes that could be observed in the present, acting gradually over an immense period of time.

Charles Lyell

Lyell presented his ideas in the influential three-volume work, *Principles of Geology*, published in the 1830s, which challenged theories about geological cataclysms proposed by proponents of catastrophism like Cuvier and Buckland. (9)

From around 1850 to 1980, most geologists endorsed uniformitarianism ('the present is the key to the past') and gradualism ('geologic change occurs slowly over long periods of time'), and rejected the idea that cataclysmic events like earthquakes, volcanic eruptions, or floods of vastly greater power than those observed at the present time, played any significant role in the formation of the Earth's surface. Instead, they believed that the Earth had been shaped by the long-term action of forces such as volcanism, earthquakes, erosion, and sedimentation, which could still be observed in action today. In part, the geologists' rejection was fostered by their impression that the catastrophists of the early 19th century believed that God was directly involved in determining the history of Earth. (10)

In the 1950s, Immanuel Velikovsky propounded catastrophism in several popular books. He speculated that the planet Venus is a former comet, which was ejected from Jupiter and subsequently - 3,500 years ago - made two catastrophic close passes by Earth, 52 years apart. It later interacted with Mars, which then had a series of near collisions with Earth, ending in 687 BC and settling into its current orbit. Velikovsky used this to explain the biblical plagues of Egypt, the biblical reference to the 'sun standing still' for a day (from the biblical book of Joshua 10:12 and 13, explained by changes in Earth's rotation), and the sinking of Atlantis. Scientists rejected Velikovsky's theories, often quite passionately. (11)

Almost 50 years later, in 1997, authors D.S. Allan and J.B. Delair published their book *Cataclysm!* Using more powerful analytical tools, they looked at the data Velikovsky examined and proposed a fresh thesis of their own.

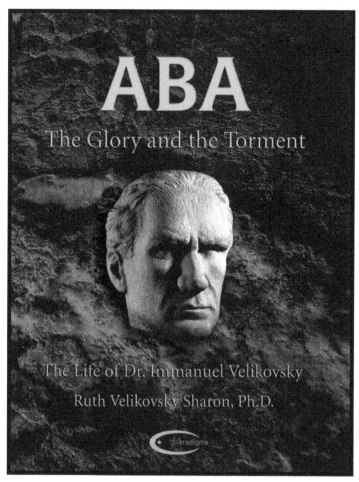

Book written about the life of Velikovsky

Like Velikovsky, they believed that folklore myths were eyewitness reports of what was seen in the sky. They described many of the myths Velikovsky described and added more that were unavailable to Velikovsky. From their research, they concluded that a fragment of an exploding supernova entered our solar system around 9500 BC. This fragment, called Phaeton-Marduk, caused one of Neptune's moons to become the planet Pluto. Phaeton-Marduk then pulled one of Saturn's moons, Chiron, away from Saturn, making it the smallest planet in the Solar System. Chiron was first discovered in 1976. Phaeton-Marduk then caused the fragmentation of the planet Taimat. Taimat's fragments are now the asteroid belts between Jupiter and Mars. Taimat's moon, Kingu, went into orbit around Phaeton-Marduk. Phaeton-Marduk then came close to the Earth disturbing the Earth's rotation. The moon, Kingu, was pulled from Taimat by Earth's gravitation. Kingu fell apart, the pieces plunging into the Earth. These combined disturbances caused the deluge. Phaeton-Marduk then flipped Venus upside down, causing it to have a backward rotation, and Phaeton-Marduk finally fell into the sun.

Allan and Delair also described geological data showing large deposits of broken bones and shattered trees mixed together in heaps, and lakes that have beds resembling craters that might have been formed by aerial bombardment of huge meteors, and many other geological abnormalities. Allan and Delair state in their book, "many had doubtless seen the great antediluvian buildings, large boats (arks included), irrigation systems, pottery, metallic utensils, and weaponry specifically mentioned in innumerable traditions they had not themselves designed." And they pose this question: *What firm evidence is there that such artificial creations existed before the Phaeton disaster?* This they date at 11,500 years ago. (13)

In this book we will explore many ancient megalithic works, especially those in Peru, Bolivia, Egypt, and Lebanon, which defy standard archaeological explanations. They could be the remnants of great highly technological civilizations that existed prior to Allan and Delair's catastrophic event.

Also, did those who have written of the end of the Ice Age as a cause of cataclysms take

Animation of a catastrophic event

on board the repeated cosmic impacts and the resulting convulsions of Earth? Did they take into account specifically the probable buckling of the Earth's surface? The possible shifting landmasses and oceans (from east/west to north/south in orientation) through the effects of the Deluge triggered by Phaeton, the fragment of Vela? The associated tilt of the globe (from vertical to the now 23.5 degrees)? And the probability that the residents of previously unknown advanced cultures, who survived the Deluge of about 12,000 years ago, aided the subsequent recovery of human societies?

Of course, the obvious candidate for an advanced civilization existing prior to the more famous ones of Egypt, Sumeria, and the Indus Valley is that of Atlantis, which will not be covered much in this book, as there are literally thousands of accounts that can be found elsewhere. However, a teaser to the idea that the Earth has not always been stable does come from the famous statements by Plato.

In the Timaeus dialogues, these being a record of discussions between the Greek Statesman Solon and an Egyptian priest, Plato reports the following:

"You Greeks are all children… you have no belief rooted in the old tradition and no knowledge hoary with age. And the reason is this. There have been and will be many different calamities to destroy mankind, the greatest of them by fire and water, and lesser ones by countless other means… You remember only one deluge, though there have been many."

Bust of Plato

What might these calamities be which Plato's Egyptian informants are referring to? Evidence has accumulated from a variety of scientific disciplines, which demonstrate that a massive cosmic object (probably a portion of an astronomically-near supernova explosion) passed close by the Earth in approximately 9500 BC. This cosmic event caused a worldwide cataclysm of enormous proportions, including massive shifting of the Earth's surface, devastating volcanic activity, mega-tsunami waves, subsidence of regional landmasses, and mass extinctions of both animals and humans. In this regard it is vitally important to note that many of the geological and biological effects previously attributed to the hypothesized glacier movements of ice age times could not have been caused by the slow movement of ice but were in fact caused by the rapid and vast displacement of oceanic bodies of water (this being caused by the irresistible gravitational pull of the enormous cosmic object passing by the Earth). Additionally, the species-wide animal extinctions caused by this event occurred far beyond the geographical boundaries set for the 'Ice Age glaciations' by orthodox theorists. (14)

Approximately 2000 years later, in roughly 7640 BC, a cometary object sped towards the Earth. This time, however, rather than passing by the Earth as the cosmic object of 9500 BC had done, the cometary object actually entered the atmosphere, broke into seven pieces, and impacted the Earth at known locations on the planet's oceans. Scientific studies of the effects of rapidly moving large objects impacting with the ocean surface have conclusively demonstrated that waves resulting from a massive cometary impact would attain vertical heights of 2 to 3 miles, with forward speeds of 400 to 500 miles per hour, and a sustained force that would carry them 2000 to 3000 miles in every direction radiating from the impact location. It is clear that these great waves would have crashed upon the shores of numerous continents, totally obliterating, especially in coastal areas of gently rising lands, all human settlements and any structures they had built.

Archaic myths from many parts of Europe (and around the world) refer to this event by mention of bright new stars which fell to Earth as seven flaming mountains, of how the oceans rose up in vast waves and totally engulfed the lands, and how summer was driven away with a cold darkness that lasted several years. In support of the mythological accounts of the vast waves covering the lands it is important to mention that many of the highest mountains in England, Scotland, and Ireland are littered with beds of sand and gravel containing sea shells deposited in the very recent geological past. Geology also gives irrefutable evidence that at two times in the recent past, around 7640 BC and 3100 BC, there have been complete reversals of the Earth's magnetic field caused by an outside influence, most probably a comet.

Estimates of the decimation of the global human population from this event range as high as 50-60% (many people would have lived on sea shores due to the availability of fish stocks). Therefore, the decimation of the planet's human population from the 9500 BC cosmic object pass-by compounded with that of the 7640 BC. cometary impacts would have severely decreased the number of humans on Earth during the following four thousand years. This is a crucial matter to consider, for the reason that orthodox archaeologists have long been mystified by both the relative scarcity of human remains from the period of 7500 to 3500 BC and, even more important, by the apparently sudden appearance of the highly developed civilizations of Megalithic Europe and Dynastic Egypt around 3100 BC.

Something akin to a global scale combustion caused by perhaps a comet scraping our planet's atmosphere or a meteorite slamming into its surface, scorched the air, melted bedrock and altered the course of Earth's history. Exactly what it was is unclear, but this event jump-started what Kenneth Tankersley, an assistant professor of anthropology and geology at the University of Cincinnati, calls the last gasp of the last ice age. "Imagine living in a time when you look outside and there are elephants walking around in Cincinnati," Tankersley says. "But by the time you're at the end of your years, there are no more elephants. It happens within your lifetime."

Tankersley explains what he and a team of international researchers found may have caused this catastrophic event in Earth's history in their research, "Evidence for Deposition of 10 Million Tonnes of Impact Spherules Across Four Continents 12,800 Years Ago," which was published in the Proceedings of the National Academy of Sciences.

This research might indicate that it wasn't the cosmic collision that extinguished the mammoths and other species, Tankersley says, but the drastic change to their environment. "The climate changed rapidly and profoundly. And coinciding with this very rapid global climate change was mass extinctions." Tankersley is an archaeological geologist. He uses geological techniques, in the field and laboratory, to solve archaeological questions. He's found a treasure trove of answers to some of those questions in Sheriden Cave in Wyandot County, Ohio. It's in that spot, 100 feet below the surface, where Tankersley has been studying geological layers that date to the Younger Dryas time period, about 13,000 years ago. About 12,000 years before the Younger Dryas, Earth was at the Last Glacial Maximum - the peak of the Ice Age. Millennia passed, and the climate began to warm. Then something happened that caused temperatures to suddenly reverse course, bringing about a century's worth of near-glacial climate that marked the start of the geologically brief Younger Dryas.

There are only about 20 generally accepted archaeological sites in the world that date to this time period, and only 12 in the United States, including Sheriden Cave.

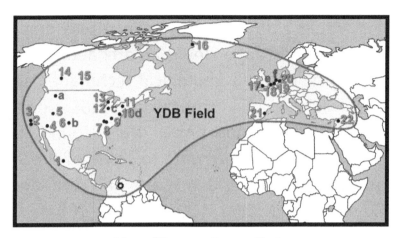

Map of Younger Dryas effect

"There aren't many places on the planet where you can actually put your finger on the end of the last ice age, and Sheriden Cave is one of those rare places where you can do that," Tankersley says. In studying this layer, Tankersley found ample evidence to support the theory that something came close enough to Earth to melt rock and produce other interesting geological phenomena. Foremost among the findings were carbon spherules. These tiny bits of carbon are formed when substances are burned at very high temperatures. The spherules exhibit characteristics that indicate their origin, whether that's from burning coal, lightning strikes, forest fires or something more extreme. Tankersley says the ones in his study could only have been formed from the combustion of rock.

The spherules also were found at 17 other sites across four continents, an estimated 10 million metric tons' worth, further supporting the idea that whatever changed Earth did so on a massive scale. It's unlikely that a wildfire or thunderstorm would leave a geological calling card that immense, covering about 50 million square kilometers. "We know something came close enough to Earth and it was hot enough that it melted rock, that's what these carbon spherules are. In order to create this type of evidence that we see around the world, it was big," Tankersley says, contrasting the effects of an event so massive with the 1883 volcanic explosion on Krakatoa in Indonesia. "When Krakatoa blew its stack, Cincinnati had no summer. Imagine winter all year round. That's just one little volcano blowing its top."

Other important findings include:

Micrometeorites - smaller pieces of meteorites or particles of cosmic dust that have made contact with Earth's surface.

Nano diamonds - microscopic diamonds formed when a carbon source is subjected to an extreme impact, often found in meteorite craters.

Lonsdaleite - a rare type of diamond, also called a hexagonal diamond, only found in non-terrestrial areas such as meteorite craters.

Tankersley says while the cosmic strike had an immediate and deadly effect, the long term side effects were far more devastating, similar to Krakatoa's aftermath but many times worse, thus making it unique in modern human history. In the cataclysm's wake, toxic gas poisoned the air and clouded the sky, causing temperatures to plummet. The roiling climate challenged the existence of plant and animal populations, and it produced what Tankersley has classified as 'winners' and 'losers' of the Younger Dryas. He says inhabitants of this time period had three choices: relocate to another environment where they could make a similar living; downsize or adjust their way of living to fit the current surroundings; or swiftly go extinct. 'Winners' chose one of the first two options, while 'losers,' such as the woolly mammoth, took the last.

Wooly mammoth

"Whatever this was, it did not cause the extinctions," Tankersley says. "Rather, this likely caused climate change. And climate change forced this scenario: You can move, downsize or you can go extinct."

Author Barbara Hand Clow examines legendary cataclysms in her 2001 book *Catastrophobia: The Truth Behind Earth Changes* and shows how, contrary to many prophecies of doom, we are actually on the cusp of an era of incredible creative growth. The recent discovery of the remains of ancient villages buried beneath the Black Sea is the latest instance of mounting evidence that many of the 'mythic' catastrophes of history - such as the fall of Atlantis and the Biblical Flood - were actual events. She shows that a series of cataclysmic disasters, caused by a massive disturbance in the Earth's crust 11,500 years ago, rocked the world and left humanity's collective psyche deeply scarred. Her inspirations for writing this book were Allan and Delair, as well as the oral traditions taught to her from her Native Cherokee grandfather. From Hand Clow's perspective, we are a wounded species, and this unprocessed fear, passed from generation to generation, is responsible for our constant expectations of apocalypse, from Y2K to the famed end of the Mayan calendar in 2012.

In her expanded edition of *Catastrophobia,* entitled *Awakening the Planetary Mind: Beyond the Trauma of the Past to a New Era of Creativity*, she discusses further the mounting geological and archaeological evidence that many of these mythic catastrophes were actual events, and further reveals the existence of a highly advanced global maritime culture that disappeared amid great Earth changes and rising seas 14,000 to 11,500 years ago, nearly causing our species' extinction. It was first published in 2011.

In her own words she states, "a recent global scientific data convergence reveals that a great cataclysm occurred only 11,500 years ago; the Late Pleistocene extinctions, according to geology, and the Flood, according to theologians. This was followed by massive crustal adjustments and flooding for thousands of years as human cultures struggled for survival while they were deeply traumatized. As this story comes forth, it emerges in a damaged world in which many people believe that the end of the world is coming soon. "

And she continues, "crippled by unnamed fear that is carried in racial memory, our surface minds are filled with floating images of disaster, guilt, and suffering. To ease our inner minds, we project these painful thoughts onto outer moving screens, which could make a coming apocalypse into a self-fulfilling prophecy. But it already happened! Based on geological, biological, paleontological, and archaeological knowledge from new dating techniques, ice-core drilling, ocean sediment cores, and computer-imaging technology, most scientists agree that a series of cataclysms occurred 14,000 to 11,500 years ago. We also know a lot about the follow-up Earth changes, such as the Black Sea Flood in 5600 BC, and the eruption of Thera on Santorini in 1600 BC. During those terrible times, our planet was afflicted with floods, erupting volcanoes, earthquakes, and massive waves of death, and we were reduced to bare survivalism. As a result of more data on cross cultural global mythology, settlement patterns, and geoarchaeology, we are achieving a global memory of our recent past. Archaeological sites come alive because we know what happened and when, and we even know a great deal about the background of the sites. Many people are afflicted with 'catastrophobia,' an intense fear of catastrophes. This new word is intended to name a psychological syndrome that causes individuals and societies to think an end is coming soon. Because they are always thinking something is coming, people are not caring for the Earth."

"Now that the date and the magnitude of the cataclysms are verified by science, we can see that it is a miracle anything survived, including ourselves. But in a way, we didn't survive, because our civilization and its cultures were obscenely obliterated. Until very recently it was thought that we have always been progressively advancing. Since this book was first published, new research, discussed in this revision, has emerged that verifies Plato's date for the fall of Atlantis, and the historical record of the fall of a previously advanced world. Since the 1980s, many researchers, most notably Graham Hancock, have been analyzing the remnants of an advanced global maritime culture from more than 12,000 years ago that vanished almost without a trace. Any evidence of such a lost world is incredibly significant. Science says we use only around 10 to 15 percent of our DNA. "I wonder if the unused DNA is the coding for a mixture of the global maritime culture's technological knowledge, psychic skills, and our emotional range that was shut down by the catastrophes. I think we must access this dormant DNA as quickly as possible, so that we can take back our role as Earth's keepers." (15)

Physicist Paul LaViolette wrote the compelling *Earth Under Fire: Humanity's Survival of the Ice Age* in 2005, which demonstrates how ancient myths and lore have preserved an accurate record of a missing era in human history, and are not simply the fantasies of cultures of the past. Compelled by his decryption of an ancient warning hidden in zodiac constellation lore, LaViolette worked with information from many scientific sources, including astronomical observations, polar ice core measurements, and other geological data, to confirm that our galaxy's core exploded, unleashing a barrage of cosmic rays that arrived near the end of the last ice age. This barrage caused the solar system to become enveloped in a dense nebula, which led to periods of persistent darkness, frigid cold, severe solar storms, searing heat, and mountainous floods that plagued mankind for many generations. Linking his scientific findings to details preserved in ancient myths and monuments, he demonstrates how past civilizations accurately recorded the causes of these cataclysmic events, from his 1983 doctoral thesis about the Galactic Superwave.

In this theory, LaViolette hypothesizes that galactic core outbursts are the most energetic phenomenon taking place in the universe. During the early 1960's, astronomers began to realize that the massive object that forms the core of a spiral or giant elliptical galaxy periodically becomes active, spewing out a fierce barrage of cosmic rays with a total energy output equal to hundreds of thousands of supernova explosions. (16)

During the 1970's, astronomers realized that the core of our own Galaxy (the Milky Way) has also had a history of recurrent outbursts, and that at periodic intervals it enters an active phase in which its rate of cosmic ray emission rises many orders of magnitude. (17) According to LaViolette, galactic core explosions actually occur about every 13,000 to 26,000 years for major outbursts and more frequently for lesser events. The emitted cosmic rays escape from the core virtually unimpeded. As they travel radially outward through the Galaxy, they form a spherical shell that advances at very close to the speed of light. The last major outburst, based on a study of astronomical and geological data, reveals that a super wave from our Galactic core impacted our solar system near the end of the last ice age, 11,000 to 16,000 years ago. (18) A more recent update suggests that the intense super waves probably occur every 12,900 years, which is half the 25,800-year cycle of the precession of the Earth's axis.

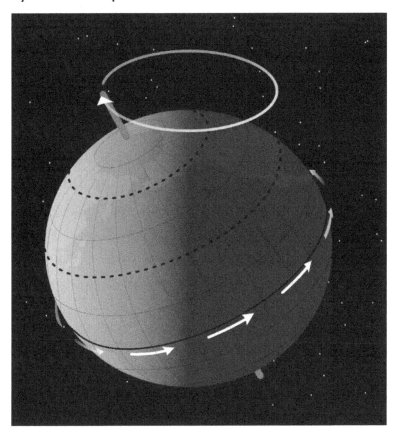

Precession or 'wobble' of the Earth

There may be a link between precession and galactic super waves. When the combination of an electromagnetic pulse, a gravity wave, radiation, and dust enter a star system, the effects on stars and planets are intense. (19)

Another main premise of the book is that these super wave events have happened to our planet and our sun countless times before, and that the last one or more events have been recorded and at least partially understood as agents of destruction. The galactic center would light up like a big blue star, bright enough to be visible even in daytime, and could have led to names like the Blue Star Kachina of the Hopi or the Eye of Ra in Egypt. Civilizations are devastated, the climate changes, magnetic poles shift, a crustal displacement or rotational pole shift may also occur, and many species die off (probably including Neanderthal man one precessional cycle, or two super waves back). Myths are created to explain all the cosmic and terrestrial phenomena observed, often as a wandering or wounded sun god, and as battles among gods who cast lightning bolts and bring down floods

Dr. Robert Schoch, professor at Boston University in the United States authored the book *Forgotten Civilization: The Role of Solar Outbursts in Our Past and Future*, published in 2012, where he briefly recaps his two decades of work on the Great Sphinx of Egypt. More importantly, he presents his latest research centered on the magnificent Göbekli Tepe complex in Turkey, which confirms his thesis that ancient civilization goes back thousands of years earlier than mainstream historians generally care to acknowledge. Also presented is a new discovery: a re-interpretation of the mysterious rongorongo texts of Easter Island, as the glyphs connect to the work of a prominent plasma physicist. He discusses how solar outbursts and plasma discharges brought about the rapid end of the last ice age and the demise of the early civilizations of that remote period. And in terms of timeframe,

Schoch believes that the last ice age ended abruptly in 9700 BC, as in 11,700 years ago, due to coronal mass ejections from the sun.

Dr. Robert Schoch in Turkey

What Allan and Delair, Hand Clow, LaViolette, and Schoch all have in common, aside from describing a major catastrophic event that happened in the past, is their placement of when this occurred, around 11,500 to 12,000 years ago. If they had differing dates, separated by thousands of years, then we would not be looking at a cohesive theory, but we are.

Whether it was a meteor or other celestial body striking the Earth, a close pass by a comet, or an energy eruption from galactic center or the sun, the result would have been devastating. Perhaps the most obvious result of this cataclysmic event, which has been well documented, is the so called Holocene extinction of large animals, especially in North America, which occurred beginning around 12,000 years ago.

These extinctions are sometimes referred to as the Quaternary extinction event, and are perhaps best known for the die off of the wooly mammoth and other mega fauna. During the last 50,000 years, including the end of the last glacial period, approximately 33 genera of large mammals have become extinct in North America. Of these, 15 genera extinctions can be reliably attributed to a brief interval of 11,500 to 10,000 radiocarbon years before present, shortly following the arrival of the Clovis people in North America. Most other extinctions are poorly constrained in time, though some definitely occurred outside of this narrow interval. (20)

Previous North American extinction pulses had occurred at the end of glaciations, but not with such an imbalance between large mammals and small ones. Moreover, previous extinction pulses were not comparable to the Quaternary extinction event; they involved primarily species replacements within ecological niches, while the latter event resulted in many ecological niches being left unoccupied.

The culture that has been connected with the wave of extinctions in North America is the Paleo-Indian culture associated with the Clovis people, who were thought to use spear throwers to kill large animals. The chief criticism of the 'prehistoric overkill hypothesis' has been that the human population at the time was too small and or not sufficiently widespread geographically to have been capable of such ecologically significant impacts. This criticism does not mean that climate change scenarios explaining the extinction are automatically to be preferred by default, however. Some form of a combination of both factors could be plausible, and overkill would be a lot easier to achieve large-scale extinction with an already dying population due to climate change.

Drawing of what some mega fauna looked like

At the end of the 19th and beginning of the 20th centuries, when scientists first realized that there had been glacial and interglacial ages, and that they were somehow associated with the prevalence or disappearance of certain animals, they surmised that the Pleistocene ice age termination might be an explanation for the extinctions.

Critics object that since there were multiple glacial advances and withdrawals in the evolutionary history of many of the mega fauna, it is rather implausible that only after the last glacial event would there be such extinctions. However, this criticism is rejected by a recent study indicating that terminal Pleistocene mega faunal community composition may have differed markedly from faunas present during earlier interglacial periods, particularly with respect to the great abundance and geographic extent of Pleistocene bison at the end of the epoch. (21)

This suggests that the survival of mega fauna populations during earlier interglacial periods is essentially irrelevant to the terminal Pleistocene extinction event, because bison were not present in similar abundance during any of the earlier interglacial periods.

The most obvious change associated with the termination of an ice age is the increase in temperature. Between 15,000 and 10,000 BC, a six-degree Celsius increase in global mean annual temperatures occurred. This was generally thought to be the cause of the extinctions. According to this hypothesis, a temperature increase sufficient to melt the Wisconsin ice sheet could have placed enough thermal stress on cold adapted mammals to cause them to die. Their heavy fur, which helps conserve body heat in the glacial cold, might have prevented the dumping of excess heat, causing the mammals to die of heat exhaustion. Large mammals, with their reduced surface area-to-volume ratio, would have fared worse than small mammals.

The catastrophic event of about 12,000 years ago, having melted much of the planet's ice and causing a global sea rise of some 350 feet, could have been so intense as to have raised global temperatures by six-degrees Celsius, as has been previously stated. This we will get into in more depth later, as we explore the idea that the ending of the last ice age was not a gradual event, as most people would assume, but fast and intense.

Finally, the Hyper Disease Hypothesis attributes the extinction of large mammals during the late Pleistocene to indirect effects of the newly arrived aboriginal humans. (22) The Hyper Disease Hypothesis proposes that humans or animals travelling with them, such as chickens or domestic dogs, introduced one or more highly virulent diseases into vulnerable populations of native mammals, eventually causing extinctions. The extinction was biased toward larger-sized species because smaller species have greater resilience due to their life history traits, as in shorter gestation time, greater population sizes, etc. Humans are thought to be the cause because other earlier immigrations of mammals into North America from Eurasia did not cause extinctions. (23)

If a disease was indeed responsible for the end-Pleistocene extinctions, then there are several criteria it must satisfy. First, the pathogen must have a stable carrier state in a reservoir species. That is, it must be able to sustain itself in the environment when there are no susceptible hosts available to infect. Second, the pathogen must have a high infection rate, such that it is able to infect virtually all individuals of all ages and sexes encountered. Third, it must be extremely lethal, with a mortality rate of around 50 to 75%. Finally, it must have the ability to infect multiple host species without posing a serious threat to humans. Humans may be infected, but the disease must not be highly lethal or able to cause an epidemic. Numerous species including wolves, mammoths, camelids, and horses had emigrated continually between Asia and North America over the past 100,000 years. For the disease hypothesis to be applicable in the case of the Americas, it would require that the population remain immunologically naive despite this constant transmission of genetic and pathogenic material.

By the 20th century, scientists had rejected old tales of world catastrophe, and were convinced that global climate could change only gradually over many tens of thousands of years. But, in the 1950s, a few scientists found evidence that some changes in the past had taken only a few thousand years. During the 1960s and 1970s other data, supported by new theories and new attitudes about human influences, reduced the time a change might require to hundreds of years. Many doubted that such a rapid shift could have befallen the planet as a whole. The 1980s and 1990s brought proof (chiefly from studies of ancient ice) that the global climate could indeed shift, radically and catastrophically, within a century, perhaps even within a decade.

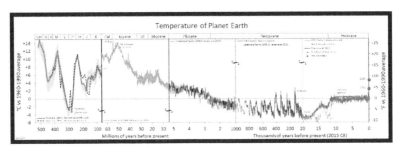

Graph showing millions of years of the Earth's surface temperature

For one group of American scientists on the ice in Greenland, the 'moment of truth' struck on a single day in midsummer 1992 as they analyzed a cylinder of ice, recently emerged from the drill hole, that came from the last years of the Younger Dryas. The Younger Dryas was a geologically brief (1,300 ± 70 years) period of cold climatic conditions and drought, which occurred between approximately 12,800 and 11,500 years BC. The American scientists saw an obvious change in the ice, visible within three snow layers that spanned scarcely three years! The team analyzing the ice was first excited, then sobered their view of how climate change had shifted irrevocably. The European team reported seeing a similar step within at most five years (later studies found a big temperature jump within a *single* year). (24) Might the change have been restricted only to parts of the world near Greenland? The first results, from the Norwegian Sea in 1992, confirmed that the abrupt changes seen in Greenland ice cores were not confined to Greenland alone. Later work on seabed cores from the California coast to the Arabian Sea, and on chemical changes recorded in cave stalagmites from Switzerland to China, confirmed that the oscillations found in the Greenland ice had been felt throughout the northern hemisphere. (25)

The Younger Dryas impact hypothesis or Clovis comet hypothesis is the theoretical large air burst or Earth impact of an object or objects from outer space that initiated the Younger Dryas cold period about 12,900 BC. (26) The hypothesized impact event scenario stated that the air burst(s) or impact(s) of a swarm of carbonaceous chondrites (nonmetallic meteorites high in carbon content), or comet fragments, set areas of the North American continent on fire, causing the extinction of most of the mega fauna and the demise of the North American Clovis culture after the last glacial period. (27) This swarm is hypothesized to have exploded above or possibly on the Laurentide Ice Sheet in the region of the Great Lakes, though no impact crater has been yet identified.

The hypothesis proposed that animal and human life in North America not directly killed by the blast or the resulting coast-to-coast wildfires, would have likely starved on the burned surface of the continent. The evidence claimed for an impact event includes a charred carbon-rich layer of soil that has been found at some 50 Clovis dated sites across the continent. The layer contains unusual materials (Nano diamonds, metallic micro spherules, carbon spherules, magnetic spherules, iridium, charcoal, soot, and fullerenes enriched in helium-3) interpreted as evidence of an impact event, at the very bottom of the black mat of organic material that marks the beginning of the Younger Dryas. (28)

The idea that Earth-based volcanism, other natural processes, or human activity being responsible have been ruled out.

Telltale points of the Clovis culture

Recent research has been reported that at Lake Cuitzeo in the central Mexican state of Guanajuato, evidence supporting the Younger Dryas impact hypothesis was found in lake bed cores dating to 12,900 BC. The evidence included properly identified Nano diamonds, carbon spherules, and magnetic spherules. Multiple analyses demonstrated the presence of three allotropes of Nano diamond: n-diamond, i-carbon, and hexagonal Nano diamond (lonsdaleite). Multiple hypotheses were examined to account for these observations, though none were believed to be terrestrial.

If it is assumed that the hypothesis supposes that all effects of the putative impact on Earth's biosphere would have been brief, all extinctions caused by the impact should have occurred simultaneously. However, there is much evidence that the mega fauna's extinctions that occurred across northern Eurasia, North and South America at the end of the Pleistocene were not synchronous. The extinctions in South America appear to have occurred at least 400 years after the extinctions in North America. (33)

If there was indeed a strike from outer space, the effect would have most likely been global, affecting wind patterns, and thus the climate of the entire planet. Whether this catastrophe was the result of celestial bodies impacting the Earth, plasma from the sun, energy from the galactic center or in fact a combination of all of the above, the effects on human populations would have most likely been as dire as those on the mega fauna.

As we look around the world, especially in Egypt, Lebanon, Turkey, the west coast of Italy, Peru, and Bolivia, there are stone structures and the remains of others which don't easily fit into the standard picture of history. The pyramids of Giza in Egypt, Puma Punku in Bolivia, and the great megalithic wall of Sachsayhuaman in Peru are but three examples of astonishingly well-made stone works which modern engineers, stone masons, and other experts puzzle over. Conventional academics in general date these structures well within the standard timeline of so-called civilization. The generally prescribed creation date of the three pyramids of Giza is about 2500 BC, Puma Punku is alleged to have been constructed around 600 AD, and Sachsayhuaman approximately 1200 to 1400 AD. However, what intrigues engineers, architects, stone masons, and other professionals is the extreme precision of the work, often in very hard stone, which many archaeologists insist was usually achieved using bronze and or copper chisels, wooden measuring devices, and stone hammers.

The rise of civilization as we know it, or are taught in school, tends to follow the following formula. Two prerequisites for civilization are the human ability to organize and the production of food in large quantities. Large amounts of food made large populations possible, but only if they could be effectively organized.

In the space of 5000 years, from 8000 to 3000 BC, the earliest settled villages grew into full civilizations in Iraq, Egypt, Anatolia, Iran, India, Pakistan, and China. Among the important steps in the movement toward civilization were irrigation, the city-state, trade, metalworking, and writing. (30) It is not accidental that the cradles of civilization were river valleys such as the Nile, Tigris, Euphrates, Indus, and Yellow.

The mysterious and famous 'H blocks' of Puma Punku

The land around these rivers must have been recognized as being rich, but the source of their richness was new soil deposited each year when the rivers flooded. The valleys were not useful to the earliest farmers until they learned to control flooding or adapt to it. The rise of civilization was partly the story of learning to control these rivers and realizing the potential of the land. More is known about the history of the Tigris, Euphrates, and Nile civilizations than others because these areas have been extensively excavated.

The Nile at Aswan

These three rivers carry water from highlands far inland to the sea, passing through very arid regions. The contrast between the land adjacent to the river and that a short distance away is striking. Desert can exist only a few hundred yards from the Nile, for example. The land around the rivers is rich, but making it bloom required the transfer of water to those parts of the valley not adjacent to the river; and the construction of large-scale irrigation projects required a large communal effort and organization. Once irrigation was understood and in place, food production soared along the rivers, making these valleys the richest and most populous places on Earth. The relative riches of the area made possible specialization of labor, leisure time, the development of the arts and, in some cases, the necessity of defense. (31) Though wars, climate shifts and other disturbances have occurred in Iraq, Egypt, Anatolia, Iran, India, Pakistan, and China, cultures there have continued to exist to the present day.

2. Peru and Bolivia

We will now discuss where megalithic remains can be found to this very day that may be examples of works done prior to the great cataclysm. Of course, this flies in the face of conventional history and archaeology, but is in fact the point of the book.

In the cases of Bolivia and Peru organized groups, or civilizations, appeared later. The Tiwanaku culture, and the area it inhabited by the same name on the southern shore of Lake Titicaca, was the largest and most famous in Bolivia. The area around Tiwanaku may have been inhabited as early as 1500 BC as a small agricultural village. (32) Around 400 AD a state in the Titicaca basin began to develop and an urban capital was built at Tiwanaku itself, which is now an archaeological site still undergoing, and awaiting, future excavation. The community grew to urban proportions between AD 600 and AD 800, becoming an important regional power in the southern Andes. Early estimates figured that the city had covered approximately 6.5 square kilometers at its maximum, with 15,000 to 30,000 inhabitants. (29)

However, satellite imaging since the late 20th century has caused researchers to dramatically raise their estimates of population. They found that the extent of fossilized suka kollus across the three primary valleys of Tiwanaku appeared to have the capacity to support a population of between 285,000 and 1,482,000 people. (30) The empire continued to grow, absorbing cultures rather than eradicating them. The elites' power continued to grow along with the surplus of resources until about 950 AD. At this time a dramatic shift in climate occurred, as is typical for the region. (31)

In Peru, though there were many cultures of relative sophistication, none rivaled the fame and levels of development of the Inca. Prior to their existence as a cohesive civilization - and some say that the Tiwanaku were their forebears - the Wari were the largest organized people of the

Andes and Peruvian coast. The Wari flourished from about 600 to 1000 AD, centered on a city called Wari near present day Ayacucho. They created new fields with terraced field technology and invested in construction of a major road network. Several centuries later, when the Inca began to expand their empire, they drew on both of these innovations. As a result of centuries of drought, the Wari culture began to deteriorate around 800 AD. Archeologists have determined that the city of Wari was dramatically depopulated by 1000 AD, although it continued to be occupied by a small number of descendant groups. Buildings in Wari and in other government centers had doorways that were deliberately blocked up, as if the Wari intended to return, someday, when the rains returned. (32)

The gradual collapse of the Wari more or less corresponded with the rise of the Inca. As was stated above, the Inca may have been the descendants of the Tiwanaku, but opinions greatly differ on this issue. It is more likely that the Inca evolved on the islands of the Sun and Moon in Lake Titicaca, and were forced out of the area by climate change, and attacks from the local Aymara people. (33) The case for the islands versus the Tiwanaku origin theory is that the terracing systems on the islands of the Sun and Moon are very Inca-like in character, as are the building construction styles. Also, Inca style pottery can be found in profusion on both islands, but the same cannot be said of Tiwanaku. Evidence of Inca ceramics at the latter was most likely from Inca occupation centuries later, and Inca-style construction was in fact far inferior to what is seen at Tiwanaku.

The Inca made Cusco their center, to the north of Lake Titicaca and it remained their capital until the arrival and destruction by the Spanish conquistadors in 1533. By this time, the Inca world stretched from Colombia in the north to the middle of Chile and Argentina in the south, the Pacific Ocean in the west and Amazon basin to the east. Some of the most astonishing achievements of the Inca, which puzzle, again, experts such as engineers and stone masons, are their constructions in stone. The arrival of the Spanish in 1532 brought knowledge of concrete, and this process was used extensively by these colonists to construct their cities, including Cusco. The Inca did not know of concrete, and used instead a local clay mortar to fill in the spaces between the stones in their buildings, walls, etc.

Astonishing accuracy of the Coricancha - the stone is basalt

One also finds, even to this day, some amazing works such as the aforementioned

Sachsayhuaman and the Coricancha in Cusco, where no mortar of any kind was used. It was stone-on-stone, with astonishing accuracy of fit. In the Inca toolkit, as found in the archaeological record, only copper and bronze chisels have been found, along with wooden measuring instruments and stone pounders or hammers. Conventional archaeologists contend that such tools were responsible for the refined workmanship seen in Cusco and other 'Inca' areas. However, the stone used - granite, andesite, and basalt - are harder than the majority of the tools used, and thus could not have been responsible for the work. The same is true of Tiwanaku and the connected site of Puma Punku. Massive megalithic blocks with sculpted surfaces are found at these locations, made of local sandstone, which would be difficult to shape with bronze chisels and stone hammers. However, the real enigmas are the even harder andesite and basalt stones, cut and shaped with such precision that modern engineers, stone masons, and other achievements of the Inca, which puzzle, again, experts such as engineers and stone masons, are their constructions in stone. The arrival of the Spanish in 1532 brought knowledge of concrete, and this process was used extensively by these colonists to construct their cities, including Cusco. The Inca did not know of concrete, and used instead a local clay mortar to fill in the spaces between the stones in their buildings, walls, etc.

Astonishing accuracy of the Coricancha - the stone is basalt

One also finds, even to this day, some amazing works such as the aforementioned

Sachsayhuaman and the Coricancha in Cusco, where no mortar of any kind was used. It was stone-on-stone, with astonishing accuracy of fit. In the Inca toolkit, as found in the archaeological record, only copper and bronze chisels have been found, along with wooden measuring instruments and stone pounders or hammers. Conventional archaeologists contend that such tools were responsible for the refined workmanship seen in Cusco and other 'Inca' areas. However, the stone used - granite, andesite, and basalt - are harder than the majority of the tools used, and thus could not have been responsible for the work. The same is true of Tiwanaku and the connected site of Puma Punku. Massive megalithic blocks with sculpted surfaces are found at these locations, made of local sandstone, which would be difficult to shape with bronze chisels and stone hammers. However, the real enigmas are the even harder andesite and basalt stones, cut and shaped with such precision that modern engineers, stone masons, and other professionals question how such work could have been achieved without at least 20th century technology.

The author with 'H blocks' at Puma Punku

Most historians and oral traditions suggest that the Inca left their homeland of the islands of the Sun and Moon, and nearby Copacabana, around 950 AD, having been driven out of the area by approximately 40 years of drought and attacks by local Aymara tribal people. Although of course Lake Titicaca has massive amounts of water, the minerals that it contains to this very day as the result of having once been ocean makes that water not useable for agriculture. It is well known that the Inca traveled to Cusco following an ancient Wari, and perhaps even earlier Tiwanaku culture trail. This trail closely hugs the Vilcanota River from its source in the high Altiplano near Pucara in present day Peru. Along this route, there is a megalithic construction at a small town called Raqchi, which was adopted by the Inca and eventually became what is known as the temple of Viracocha.

Megalithic foundations below adobe work at Temple of Viracocha

Viracocha could have been the creator God of the Inca, a separate and earlier race known as the Viracochan, or the high Sapa ruling Inca that named himself after the creator. The evidence that the Inca inherited an older and far more technologically sophisticated site is most evident in the central wall of the temple of the Viracocha complex. Here we can see that the base of the construction is made up of large and tight fitting basalt blocks whereas above is rough stone and mud clay work. Since the lower work must be older, it is safe to presume that it is pre-Inca.

As stone is not organic in nature, it cannot be dated using a technique such as radiocarbon testing, yet some new techniques like cosmogenic testing are being developed which may allow accurate dating in the future. The fact that the original works only go about a meter or so above the ground could either indicate that it was partially deconstructed, or perhaps the victim of ancient catastrophic damage.

Following the trail that the Inca would have taken in order to find or establish Cusco as their new homeland, which again closely follows the Vilcanota River, takes a strange left turn about 50 kilometers before the city itself. The river continues on its natural course into the Sacred Valley of Peru, and then continues in a continued northerly direction past Machu Pic'chu, eventually joining the Ucayali River in the Peruvian jungle and then the Amazon. Instead, the trail leads straight through another megalithic work, called the Inti Punku or sun gate.

Inti Punku showing obvious megalithic elements

This structure again has two distinct building styles - two massive walls of tight fitting basalt blocks, and then mud clay built with local stone. It would appear that here, like at the temple of Viracocha, the Inca discovered a very ancient megalithic work and constructed on top and around it. The original stone works are so weathered that there are no apparent tool marks, on the surface at least, and cracks in the blocks would seem to show evidence of a catastrophic disaster having hit it, rather than simply time and gravity.

As the trail continues, and is in fact now the main highway from the south entering modern Cusco, it takes us right to the core of ancient Inca Cusco itself. It is here that the author believes that rather than finding virgin ground in which to establish their new home, the Inca found the ruins of an abandoned and destroyed megalithic city. The evidence, which most archaeologists and other academics refuse to accept or in fact do not even see, is obvious to even the most slightly aware observer.

Starting at the center of the Inca world, the Coricancha (courtyard of gold), which was converted into a Catholic church as soon as the Spanish arrived in 1533, we see ample evidence of very advanced stone construction techniques well beyond the capacity of the Inca. The majority of the ancient stone work of the Coricancha, which could very well be the finest stone-on-stone mortar free construction in all of the Americas, are basalt, gleaned from a quarry 50 kilometers from the Coricancha.

Astonishing original stonework of the Coricancha (on the right)

In most cases the walls are one meter thick and the stone-on-stone contact is flawless from the outside to the inside, a feat that bewilders modern engineers and stone masons alike.

That the Inca could not have made it with bronze tools is obvious, and the first ever topographical map of Cusco, made soon after the Spanish entered and claimed the city, shows that the Inca constructions are indicated by thin lines, while the megalithic ones are thicker lines. This drawing shows us that the Inca in fact repaired the Coricancha, especially the southern wall.

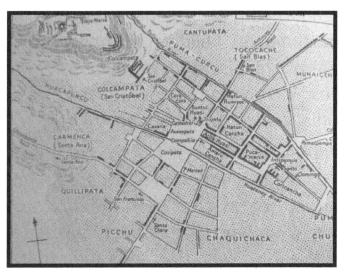

Original colonial map of Cusco showing megalithic walls

To the author, this again indicates that the Inca did not find the Coricancha in pristine condition, but a damaged structure that must have been hit by at least one massive cataclysm.

A few blocks to the northeast is what is commonly called the Inca Roca wall, named after the high Sapa Inca, number 6 of 12 in line of succession that had his palace inside a court made of green granitic stone. This is one of the classic examples of 'Inca' polygonal masonry, and the source of all of the green stone that once made up this courtyard came from a nearby neighborhood called San Blas. Unfortunately, building of houses during Spanish colonial and perhaps even Inca times has obscured the quarry, as the buildings are now right on top of it.

A very early photo of the southern wall of the complex shows that a more standard Inca wall of smaller local andesite stones along with clay mortar once covered the entire earlier wall, thus indicating that the green megalithic work is older.

Inca Roca wall with amazing tight joinery

Why the Inca decided to cover over the earlier work is unknown, but it could be that because the ancient wall had been badly damaged by one or more cataclysmic attacks, the Inca filled in gaps and wanted a more homogenous look. In the 1950s, archaeologists removed the Inca 'cover' wall to expose all of the earlier megalithic works. What is also quite curious is that some of the surface on the north wall appear to have been heat scorched, and this could lead credence to Dr. Robert

Schoch's idea that plasma from the sun struck certain parts of the Earth, especially high elevations like the city of Cusco.

What is also apparent is repair work having been done by perhaps the Inca, or perhaps a more advanced culture, on the northeast corner of the courtyard. In the photos, you can clearly see that smaller basalt blocks have been inserted into the polygonal wall at the top, glaringly different in color to the green granitic stone, which makes up most of the courtyard.

Inca Roca wall section showing possible plasma damage

Inca Roca was the 6th of the great Inca of 12. Curiously, we can see in the constructions of Cusco that the first six Inca each had their own courtyard, which was megalithic in nature. After the ancient aspects of pre-Inca Cusco were used up, then the heirs had to build their own courtyards of smaller local stones, recycled megalithic pieces, and clay mud used as mortar.

Inca Roca wall showing inferior repairs above

The reason why each ruler had to have his own courtyard was because Inca inheritance did not include the father's palace. A ruling Inca had what was called his Panaca, or ruling family, who would live inside the confines of his courtyard. Once the ruler died his Panaca, which included his high queen, or Qoya, younger children, and other family members, was replaced by the next high Inca who would choose his own people to be in his Panaca – essentially his government. This new Panaca would have to have its own palace inside a courtyard. This is especially emphasized by perhaps most famous of the Sapa Inca, called Pachacutec ('he who transforms or turns over the world').

Inca wall with inserted megalithic basalt recycled stones

It is known that he was born in his father's palace, and the photos show no megalithic elements to its construction - all is Inca work of smallish stones and clay mortar.

Since parts of the wall are deconstructed, likely by the Spanish in order to build their own constructions during colonial times, we can see how the Inca masons 'cheated.' They made the fronts of the stones fit together nicely, perhaps in order to make the look blend in with the more ancient works, but in behind the joints taper back, showing a pie-shaped gap.

Inca stone work with pie slice-shaped gap

From here we head to the first established Sapa Inca's palace of Manco Capac (or Mallku Capac) located on a hill just north of the main square of Cusco. The majority of his palace was destroyed by the Spanish in order to build colonial Cusco, and perhaps as a way to demoralize the Inca descendants as well. What we find here are some of the most beautiful of true Inca wall construction; andesite stones shaped with stone and meteorite hammers, which the Inca did possess. The lintels above each trapezoid niche are basalt, again from the quarry 50 kilometers to the south, each finely hewn.

In behind the wall are the remains of an amazingly built wall section, now in ruins, that again looks like it suffered catastrophic damage rather than Spanish desecration. What seems apparent to the author is that Manco Capac, upon his entrance into Cusco some 1000 years ago, chose this site for his palace so that he would have a perfect view of the city.

Manco Capac wall section with basalt lintel

There is also evidence that tunnels exist under this location, connecting the massive megalithic location called Sacsayhuaman above, with the Coricancha below. This tunnel or system of tunnels were certainly not made by the Inca, and could be part of the more ancient megalithic works, or perhaps enhanced natural phenomena.

Sacsayhuaman is a truly megalithic site in that its main wall contains stones as large as 125 tons. The limestone quarry from which the rock was extracted is believed to only be 3 kilometers, but the rolling landscape

Possible entrance to the ancient tunnel system of Cusco

itself would make it difficult even in modern times to move such massive things. The author visited the quarry in 2015 with local experts, but could not find any evidence of tool marks on the stone surfaces. This would lead one to conclude that either all of the stones were found in situ, or that the site is so old that clearly it was not a creation of the Inca. Curiously, the surfaces of the stones do not bear any tool marks whatsoever, except bruise marks that were the result of some Inca period reshaping, or reconstruction by archaeologists in the 20th century.

The probable limestone quarry for the Sacsayhuaman stones

The vast majority of academics are completely convinced that Sacsayhuaman, in total was a creation of the Inca, but cannot explain how the Inca could have cut the stones from the bedrock, moved them, and then finally put them together with almost surgical precision. Their evidence, such as it is, is that all of Cusco and surrounding area constructions were the work of the Inca, as much more primitive cultures pre-existed them. They also depend on accounts written down by the Spanish conquistadors, which are accounts (or more likely tortured confessions) from the Inca people themselves. What is more likely is that the Inca claimed that they built at Sacsayhuaman, not that they constructed the entire complex.

Oral tradition accounts that the author was able to glean from living local experts are that the Spanish, upon first seeing the great wall of Sacsayhuaman in 1533, and shocked at its scale, asked the local Inca if their ancestors had built it.

The author and guests at Sachsayhuaman

reconstruction by archaeologists in the 20th century.

The probable limestone quarry for the Sacsayhuaman stones

The vast majority of academics are completely convinced that Sacsayhuaman, in total was a creation of the Inca, but cannot explain how the Inca could have cut the stones from the bedrock, moved them, and then finally put them together with almost surgical precision. Their evidence, such as it is, is that all of Cusco and surrounding area constructions were the work of the Inca, as much more primitive cultures pre-existed them. They also depend on accounts written down by the Spanish conquistadors, which are accounts (or more likely tortured confessions) from the Inca people themselves. What is more likely is that the Inca claimed that they built at Sacsayhuaman, not that they constructed the entire complex.

Oral tradition accounts that the author was able to glean from living local experts are that the Spanish, upon first seeing the great wall of Sacsayhuaman in 1533, and shocked at its scale, asked the local Inca if their ancestors had built it.

The author and guests at Sachsayhuaman

The answer was a simple, "No," that the wall was there when the Inca arrived about 500 years prior. Sacsayhuaman is a huge and complex site, and it is clear that varying techniques were used in its construction. We can see many remains of actual Inca construction techniques, using the local andesite stone, as well as reconstructions that the Inca had done using, in some cases, a mix of andesite and basalt blocks from the Rumicolca quarry some 50 kilometers away. Then of course there are the huge limestone pieces from the quarry 3 kilometers in the distance.

Megalithic work and smaller Inca repairs at Sachsayhuaman

There is no practical reason why such huge blocks were required at Sacsayhuaman, and they are in fact the largest ever employed in the Americas. The common belief that it was a fortress of some kind constructed by the Inca is an idea that the Spanish came up with. For the Inca, it was in fact one of the most sacred of holy places.

Moving farther afield, Pisaq was a major Inca complex located on top of a mountain near the southern end of the Sacred Valley in Peru, near Cusco. It, like Sacsayhuaman is a very large complex, and once had a spiritual sector, royal quarters, army barracks, administrative center, and vast Andene terraces with housing for the farmers. Most academics believe it was the precursor to the more famous Machu Pic'chu, constructed possibly in the 14th century. However, like the other places we have looked at so, Pisaq has megalithic elements which the Inca would have been very hard pressed to have achieved.

What first catches one's eye is a damaged megalithic wall at the base of the administrative area. The stones here are very tight fitting with no mortar, show no tool marks, and the stone appears to be basalt. If this is the case then the stone likely came from the Rumicolca quarry, which is about 30 kilometers away.

Basalt megalithic wall at Pisaq

At the spiritual center there are many damaged constructions seemingly made of basalt, right next to much poorer works of andesite and clay mud. One could very well infer that the Inca found a much earlier site of mainly basalt construction, and then added their own buildings - a common trait of many places in the area.

Traveling north in the Sacred Valley we do encounter many Inca sites that are pure Inca in design and execution.

Pisaq spiritual center - note the walls on the right

Of note is the winter palace of the brother of Huayna Capac, who was the 10th high Inca, living in the late 15th and early 16th centuries AD. As the photos show, it is of adobe clay construction with a core of field and river stone, cemented together with more adobe. As he was the high ruler of the Inca civilization, he would have at his disposal the finest stone masons in the land. Thus, the fact that his own palace was made of adobe and field/river stone adds more fuel to the idea that the fine megalithic works were not achieved by the Inca people.

Royal Inca work of the late 15th century

It is also possible that this work dates back to Pachacutec Inka Yupanqui, who was the ninth Sapa Inca (ruling from 1438 to 1471/1472) of the Kingdom of Cusco. which he supposedly transformed into a much expanded federation of states. In total, it is believed that there were 12 Inca high leaders in succession, and thus he would have lived when the technical prowess of the Inca people was close to its zenith.

Most of the high Inca rulers had a winter palace in the Sacred Valley, as the land there is about 700 meters below Cusco, and also a year-round spring/summer climate. None that the author has visited have any megalithic elements, and thus were created 100 percent by the Inca. It is when we get to Ollantaytambo, at the northern end of the Sacred Valley, that we see extreme examples of differences between Inca works and those of the much older megalithic builders.

Ollantaytambo is another vast ancient site, and has some examples of the finest of Inca period Andene terraces, some being at least four meters high or more. The method of construction is typical Inca: field stone adhered with clay mud as mortar. Even the structure dedicated to the Virgins of the Sun, who were the female priests responsible for making the royal Inca family's clothing, and preparers of their food, is an adobe and rough stone construction.

Megalithic works in back and above; Inca terraces on the right

Clearly, if the Inca stonemasons had the capability to make the megalithic works, then the Virgins of the Sun's temple would have been made that way.

The staircases that lead up to the higher levels at Ollantaytambo were honestly quite poorly constructed, with uneven heights. In some cases, the steps were made of a solid piece of reddish/purple granite, while other

Adobe and rough stone construction at Ollantaytambo

steps were of multiple pieces of stone, some local and others not. And yet when you reach the proposed Sun Temple, you find six massive slabs of the red granite that fit tightly together to this very day. The local stone is a type of fractious slate, and that is what the entire hillside is composed of. Much of the Inca terracing system was constructed using this, clay mortar, and broken pieces of the red granite.

The granite, astonishingly, comes from the top face of a mountain on the other side of

Six of the massive slabs of granite at Ollantaytambo - he quarry is the mountain to the left

the Sacred Valley. At least one of the six massive slabs is estimated to weigh 67 tons, and obviously would have originally weighed much more when it was cut from the mountainside. The question is, how was it and the others cut and transported? There is a road that leads to the quarry from the valley floor, but the upper third is not wide enough to handle these massive slabs. There is also another shorter wall to the south of the large slab one.

Most academics believe that the Inca were constructing the Temple of the Sun and then mysteriously abandoned the project, no reason given. It seems far more likely to the author that the so-called temple was a much earlier construction that was destroyed not by the Spanish, nor the Inca, but an ancient cataclysm. All of the other sites in Peru that we have explored so far show the same level of damage. It is possible that at one time the now ruined granite temple was intact.

Clear evidence that the Inca made additions to a damaged megalithic wall

The great cataclysm could have been so immense that it caused the sides to fall down to the valley floor below, breaking once great megalithic blocks into pieces. This would explain why some of the stairs were made of a solid piece - the Inca simply recycled the remains where they could.

Since the front and back walls are still partially intact and were built into the mountainside, it is possible that the cataclysm spared them due to the reinforcement aspects that were part of the original construction. As well, there is a so-called ramp that leads from the side of the temple down to the valley floor. Most academics believe that this was made by the Inca as a way to move the massive granite slabs up to the temple site. However, there is an Inca period wall that goes right across it. So, when did the Inca abandon the construction of the temple? I am sure you can answer that. Next we explore Machu Pic'chu, which definitely has some aspects that academics cannot - or will not -answer.

It is generally believed that Machu Pic'chu was constructed by the high Sapa Inca

Pachacutec in the mid-15th century as a place of rest for himself and the royal Inca family. Most estimates are that the entire complex took from 30 to 50 years to build, using at least 5000 workers to accomplish the task. The finest stone masons would likely have been brought in from Cusco, some 75 or more kilometers away, as well as most of the rest of the construction crew. The problem with this idea is, where would such a crew have lived? Who would have fed them? And what?

Hiram Bingham III, the American explorer who was backed by The National Geographic and Yale University, and was the first foreigner to make Machu Pic'chu famous in an entire issue dedicated to the site in April 1913, felt that Machu Pic'chu must have taken hundreds or thousands of years to construct.

Machu Pic'chu was a secret installation and the general public of the Inca would have had no clue that it existed. This was likely done on purpose, as rather than a place for the Inca nobility to drink and cavort, Machu

Pic'chu was more likely similar to the United States' Camp David - a place where high officials could meet in private.

Photo showing the sheer scale of Machu Pic'chu

While the site was fully active prior to the arrival of the Spanish to Cusco in 1533, the Inca world, known as the Tawantinsuyu (the four quarters of the world) stretched from southern Colombia in the north, to the center of Argentina and Chile in the south, the Pacific Ocean to the west, and the Amazon basin in the east.

Such a vast confederation of states (not an empire as most writers suggest) had many officials that governed many regions on behalf of the Inca royalty. In order for cohesive knowledge of what was going on as regards agricultural production, local disturbances, mining, and other activities in each of these areas, officials would have to occasionally visit Cusco in order to deliver this information to the Inca. As the Inca themselves strictly controlled the flow of knowledge and affairs of state, discussions were held in Machu Pic'chu, especially if they were of a sensitive nature.

There were two official entrances into the site, which were heavily guarded and both could easily be sealed off if required. Once Cusco was invaded in 1533 by the Spanish, word was sent along the Inca roads and trails to Machu Pic'chu by the highly efficient Chasqui runners and efforts were soon set into play to abandon the site. The road systems were destroyed, and that is why the Spanish never found Machu Pic'chu. It also helped that the local people either never knew of its existence, or were sympathetic to the plight of the Inca and thus kept their mouths shut when the Spanish, or their descendants, inquired about Inca places they were not aware of.

This meant that Machu Pic'chu lay dormant from about 1533 to 1911, and though time took a toll on the buildings and other structures there, very little reconstruction has occurred. This allows us to explore the site, and carefully examine how it was built. The most shocking aspect of such an exploration, which the author has now done on 50 separate occasions, is the variation in the quality of workmanship. Although most scholars insist that the finest workers were dedicated to the palatial and spiritual buildings, and that lesser important structures were not given major design and construction considerations, the following photos will show you that such a theory makes no sense whatsoever.

On the right of the following photo is what is known as the Temple of the Sun, which was at least a solar observatory, if not more. The stones fit together with an astonishing level of precision, with no mortar being used.

The work on the left and the right are profoundly different

On the left, right next to the temple, you can see that the workmanship is profoundly poorer, with roughly broken stones being cemented together with local clay as mortar. The idea that such a massive place as Machu Pic'chu would have varying degrees of quality construction is understandable, but not side-by-side as you see in the above photo. All of the stone at Machu Pic'chu is white granite, which is hard, as it contains quartz. The bronze Inca tools would not have been able to create the precision of the Temple of the Sun.

Somewhat near the Temple of the Sun, in what appears to be the megalithic core of Machu Pic'chu (that does include the aforementioned temple) we have what Hiram Bingham named the Temple of the Three Windows, in the photo below. What you can clearly see is that the walls in front and to the left are of poor construction technique, while that in the back is of huge stones that once fit together almost perfectly.

Extreme differences in construction techniques

The large gaps have been explained by geologists, who have been with the author, as being clear signs of massive earthquake activity and possibly even that of the ancient cataclysm of about 12,000 years ago. If this is the case, then Inca clearly found an ancient abandoned city, adopted it, and built around the ancient megalithic works.

Megalithic below and Inca above

The above photo shows the other end of the Temple of the Three Windows, and here you can clearly see that the work above is

tools would not have been able to create the precision of the Temple of the Sun.

Somewhat near the Temple of the Sun, in what appears to be the megalithic core of Machu Pic'chu (that does include the aforementioned temple) we have what Hiram Bingham named the Temple of the Three Windows, in the photo below. What you can clearly see is that the walls in front and to the left are of poor construction technique, while that in the back is of huge stones that once fit together almost perfectly.

Extreme differences in construction techniques

The large gaps have been explained by geologists, who have been with the author, as being clear signs of massive earthquake activity and possibly even that of the ancient cataclysm of about 12,000 years ago. If this is the case, then Inca clearly found an ancient abandoned city, adopted it, and built around the ancient megalithic works.

Megalithic below and Inca above

The above photo shows the other end of the Temple of the Three Windows, and here you can clearly see that the work above is far inferior to that below. As this wall was not reconstructed, it appears to be a clear example, again, that the Inca found a damaged megalithic structure and added their work above it. If there were two or five examples of this, you could say that perhaps it was coincidence. However, there are far more examples than that, best seen in person, or in one of my YouTube videos.

Many of the other megalithic works at Machu Pic'chu show large gaps as in the two previous photos. The width of these gaps is very similar, and again suggest that a massive earthquake cataclysm, moving in an east-west direction, affected the site in the very distant past.

More evidence of catastrophic damage can be seen in the next photo. The right wall has sunken down almost a meter lower than the back wall, and that on the left. A geologist who inspected this with the author stated quite emphatically that such a drop would not have been the result of poor foundation work when the structure was made, but far more likely, again, the result of a catastrophic earthquake. She also stated that such an earthquake would have caused the rough stone and clay mortar buildings to be completely demolished. This, then, clearly indicates that the megalithic core of Machu Pic'chu, which comprises about 5-10 percent, is older than the Inca, and may in fact have been made prior to the cataclysmic event of about 12,000 years ago.

Wall showing extreme catastrophic damage

All of the famous ancient sites in the Cusco area, including Sacsayhuaman, Machu Pic'chu, Pisaq, Ollantaytambo, Qenqo, and many others all show signs of catastrophic ancient damage, and later Inca repair work.

The last place we will explore, though we could go on for several pages, is Saywite, located about four hours drive from Cusco.

The famous carved stone at Saywite

It is most famous for a large stone with various animal, geometric, and stair shapes carved into it. Unfortunately, perhaps only 10 to 20 visitors per day see it. What practically no one aside from locals knows is that there are many shaped stones and temples located just below the stone in a small valley. Clearly the temples were made by the Inca, but some of the megalithic stones in the area are broken in half, not the result of Spanish colonial quarrying, but cataclysmic in nature. Furthermore, Saywite, being a four-hour drive from Cusco, would hardly be a place where one would quarry stone, when there are many other ancient places much closer.

Quad copter view of one of the ancient Saywite stones

Another view of one of the broken stones of Saywite

Bolivia does not have the wealth of historical artifacts that Peru has, and the most famous pre-Colombian location is that of Tiwanaku and Puma Punku, just south of Lake Titicaca. Conventional academics believe that the major works at Tiwanaku and Puma Punku, which are in fact the same place separated by fences, were constructed by the Tiwanaku culture between 500 and 900 AD. The site was then abandoned due to a 40-year drought and invasions by local Aymara tribes, who occupy these lands to this very day.

Classic view of a restored part of Tiwanaku

All of the stone work involved in the constructions at Tiwanaku, in general, is either red sandstone from a quarry located 10 kilometers to the south, or grey andesite taken from Cerro Khapia, a volcano 70 kilometers to the north. These are extreme distances to consider, especially since one slab of the red sandstone weighs an estimated 131 tons. Also, it is believed that at least 90 percent of Tiwanaku and Puma Punku has been systematically removed to locals from nearby towns and villages, as well as the capital La Paz, over the course of the last 1000 years. Thus, there could have been much larger stones there at one time.

Inside the crater of Cerro Khapia

The question of how the stones from both quarries were moved to the Tiwanaku and Puma Punku site has been addressed by academics, with poor to ridiculous answers. Some say that the red sandstone was moved using wooden rollers, as in tree trunks. The problem with this idea is that Tiwanaku and Puma Punku are above the tree line. The other is that the andesite from Cerro Khapia was moved using boats made of the local totora reed, which is a ridiculous presumption.

The precision of many of the flat surfaces is astonishing. In some cases, they are almost as flat as laser perfection, and the idea that a Bronze Age culture like the Tiwanaku were responsible for this work is clearly impossible.

What is also curious is that much of the stone has been partially or fully excavated from the red clay mud of the area, which infers either extreme age, or that a cataclysmic event occurred here, partially burying the site. Evidence of this can be seen in the next photo.

Just a sample of the precision surfaces at Puma Punku

Further, there are blocks which appear to have been snapped in half - not by the invading Aymara, colonial Spanish, or more recently, but at a time in the distant past. The logic behind this statement is that there are no apparent tool marks or other evidence of attempts to break the stone.

We can also see that Kalasasaya pyramid at Tiwanaku was buried by at least two meters or more of mud. This would not be the result of slow sedimentation over the course of a long time, but a massive catastrophic event. In fact, the entire area, including the nearby town of Tiwanaku and Puma Punku, appear to have very thick deposits of the same red clay that enveloped the Kalasasaya in the past.

Partial excavation at Puma Punku

Wall of mud that struck Tiwanaku

Massive snapped block and other damaged megaliths

There is also a restaurant just west of the Puma Punku site, on private property. Here the owner, who is also an archaeologist, made a discovery when digging for a foundation - an ancient cemetery.

However, whereas a normal cemetery contains bodies individually and carefully buried, most of the human remains and ceramics in this site look shattered, as if they were buried en masse due to a cataclysmic event.

Proposed cemetery near Puma Punku

As with what we have seen in Peru, the Tiwanaku could not have made the precise surfaces, quarried or moved the massive stones of Tiwanaku and Puma Punku. Thus, it is clear that an older culture did. Also, at Tiwanaku and Puma Punku we see clear evidence of cataclysmic damage that preceded the Tiwanaku people, just as we did with the Inca culture.

And now, on to Egypt...

3. Egypt

The author with the Khemit School in 2015

The most famous of all ancient civilizations actively studied to this day must be Egypt. Egyptian civilization coalesced around 3150 BC according to conventional Egyptian chronology, with the political unification of Upper and Lower Egypt under the first pharaoh. The history of ancient Egypt includes a series of stable Kingdoms, separated by periods of relative instability known as Intermediate Periods: the Old

Kingdom of the Early Bronze Age, the Middle Kingdom of the Middle Bronze Age, and the New Kingdom of the Late Bronze Age.

Note that all three of the famous Kingdoms, during which most Egyptologists believe that the Sphinx, three great Giza pyramids, and other famous and lesser known structures were built, were all Bronze Age cultures. More specifically, the above mentioned works are theorized to have all been constructed during the Old Kingdom (2686 to 2181 BC.) (34)

However, though the Sphinx itself, and many of the pyramids of Giza, are largely comprised of relatively soft limestone, rose granite and other harder stones were used in the inner core of the pyramids, and the casing stone to some degree. This means that the Egyptian culture, like Peru and Bolivia, is thought to have been capable of working with very hard stone like granite, basalt, and diorite - successfully and with amazing precision prior to an iron age.

One of many massive granite boxes underground at Saqqara

Iron is a very common element and iron ores occur in the mountainous areas of the eastern desert and Sinai of Egypt, though high-grade ores are rare. That and the lack of hardwood or coal needed to achieve high temperatures prevented any large-scale iron production in Egypt. (35) Native iron of meteoric origin with a high nickel content was the first metallic iron to be used during the pre-dynastic, but seems to have been limited to the making of beads and other small personal adornments. During the New Kingdom and the Third Intermediate Period, no or little iron was produced locally, and finds are few. In the 7th century BC, Ionians began to settle in the Delta and seem to have brought with them the knowledge necessary for working iron. Naukratis and Defenneh became the great Egyptian centers where iron tools were manufactured. Once they got going it took only about a century for the production of iron implements to equal the manufacture of bronze tools and weapons. (36)

What this is telling us is that iron, let alone steel or even harder materials used today to cut and shape hard stone such as granite, was not in common use in Egypt until at least 1500 years *after* the pyramids of Giza were constructed. Further, in Peru as well as Bolivia, iron and steel were not known to any extent by the people of those lands until the Spanish arrived in the 16th century AD.

Thanks to explorations with the Khemit School (www.khemitology.com), located across the street from the Sphinx entrance to the Giza plateau, on four yearly occasions so far, the author has seen ample evidence of the existence of complex tool marks in hard stone surfaces such as granite and basalt, as well as softer materials like limestone, in the Giza area and other famous and lesser known locations. The more obvious tools in use that left their marks behind were circular saws and drills, though routers and other forms of cutting technology could have been used.

Examples of dynastic Egyptian tools

Let us start our exploration in the Cairo Museum, which up until late 2015 did not allow photos to be taken of their thousands of artifacts. The collection on display is only a small percentage of that which they possess, likely as the result of lack of space. However, some believe that they are in possession of many artifacts that could have been parts of ancient machines and do not display them because they do not fit with the historical paradigm that they desperately want to protect.

The Egyptian Museum of Antiquities contains many important pieces of ancient Egyptian history. It houses the world's largest collection of Pharaonic antiquities. The Government of Egypt established the museum, built in 1835, near the Ezbekeyah Garden and later moved to the Cairo Citadel. In 1855, Archduke Maximilian of Austria was given all of the artifacts by the Government of Egypt and these are now in the Kunsthistorisches Museum in Vienna. A new museum was established at Boulaq in 1858 in a former warehouse, following the foundation of the new Antiquities Department under the direction of Auguste Mariette. The building lay on the bank of the Nile River, and in 1878 it suffered significant damage in a flood. In 1891, the collections were moved to a former royal palace in the Giza district of Cairo. They remained there until 1902 when they were moved, for the last time, to the current museum in Tahir Square.

Two large boxes outside of the Cairo Museum

In the photo above, we see two large boxes located out in front of the entrance to the museum. The one on the right is dynastic Egyptian, and this is known because the inscriptions in the ancient Egyptian language can be read. The one on the left, however, is larger, more precisely shaped, and is labeled pre-Dynastic. There are also possess, likely as the result of lack of space. However, some believe that they are in possession of many artifacts that could have been parts of ancient machines and do not display them because they do not fit with the historical paradigm that they desperately want to protect.

The Egyptian Museum of Antiquities contains many important pieces of ancient Egyptian history. It houses the world's largest collection of Pharaonic antiquities. The Government of Egypt established the museum, built in 1835, near the Ezbekeyah Garden and later moved to the Cairo Citadel. In 1855, Archduke Maximilian of Austria was given all of the artifacts by the Government of Egypt and these are now in the Kunsthistorisches Museum in Vienna. A new museum was established at Boulaq in 1858 in a former warehouse, following the foundation of the new Antiquities Department under the direction of Auguste Mariette. The building lay on the bank of the Nile River, and in 1878 it suffered significant damage in a flood. In 1891, the collections were moved to a former royal palace in the Giza district of Cairo. They remained there until 1902 when they were moved, for the last time, to the current museum in Tahir Square.

Two large boxes outside of the Cairo Museum

In the photo above, we see two large boxes located out in front of the entrance to the museum. The one on the right is dynastic Egyptian, and this is known because the inscriptions in the ancient Egyptian language can be read. The one on the left, however, is larger, more precisely shaped, and is labeled pre-Dynastic. There are also no inscriptions whatsoever on or in the box, or on the lid. This simple photograph by itself is telling us that whoever made the larger box on the left was technically more proficient than the Dynastic workers. We are very fortunate in that on our yearly expedition, we have had a Canadian geologist with us. While both of these boxes are listed as being granite, and the presumed location of the quarry from which the stone was harvested is stated by Egyptologists to be from the Aswan quarry some 500 miles to the south, our geologist believes it may in fact come other quarries in the Sinai area of eastern Egypt.

If this proves to be the case, then while academics believe that Aswan stone could have been floated down the Nile on rafts or boats, how would such large pieces be transported across the desert?

The granite of Aswan tends be very large and coarse-grained, while the box on the left was made from a much smaller grained type of granite called syenite, or possibly diorite, which does not appear to have geologically formed in the Aswan area, but does seem to exist in the Sinai area, and possibly at the place called Wadi Hammamat, which is east of Luxor.

Wadi Hammamat is located about half way between Qusier and Gift (ancient Coptos), and is famous today mostly for its pharaonic graffiti. More than 200 hieroglyphic tablets adorn the quarries of the renowned 'bekhen' stone, which is actually made up of three distinct materials. However, the graffiti transverses time and extends into the 20th century and the reign of King Farouk. (37)

Inside the museum itself, on the main floor and in a corner alcove is a box that was never completed. As the three following photos show, someone was attempting to cut off a large slab from the bottom in order to likely make the lid. The saws that were being used went off course, causing half of the slab to snap off, and the project was then apparently abandoned.

Obvious saw cut in a box inside the museum

Members of the Khemit School tour inspecting the saw mark

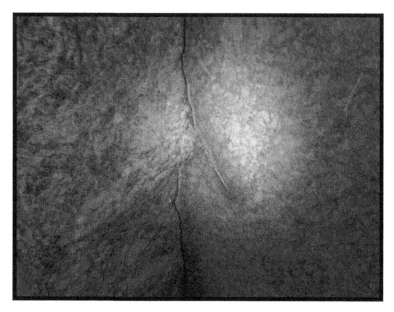
Circular saw marks in the box inside the museum

The above picture clearly shows that two circular saws were at work, one from the top and another from the bottom. They were not perfectly aligned but were cutting through the granite stone very efficiently. The only saws we have in modern times that can do such work have diamond abrasives imbedded in either high carbon or cobalt steel blades, powered by very strong electric or petroleum powered engines. As the dynastic Egyptians for most of their history had at best bronze tools, and there is no evidence of them having circular saws, they could not have done this work.

Also of note is that the interior of the box had been almost finished, but not completely, at or before the time of the broken lid incident. What is curious is that the interior corners are not crisp, but rounded slightly, indicating that it was not a saw that cut out the interior.

View of the box showing the interior

And yet, whatever form of technology was used, it left an even pebbly surface such you get with using a sandblaster.

The boxes mentioned so far are listed as sarcophagi by Egyptologists, but since the dynastic people were incapable of shaping them in the first place, their actual original function remains unknown.

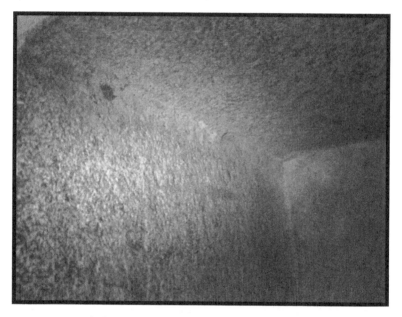

Slightly rounded corners inside the box

The Egyptians could have used them for funerary function, but they did not create them. These are the first indications that many artifacts of ancient Egypt were in fact found by the dynasties when they entered these lands around 5000 years ago, as we shall see.

Another large box with crude etching on the outside, inferior to the box itself

Friend Hans inspecting flat saw marks on another lid

Possibly the strangest and most inexplicable artifact on display is what is commonly known as the 'schist disk.' It is a tri-lobed circular object made of stone, and the label in the museum states that it could have been a "vase for lotuses." However, anyone with any sense of function or engineering would quickly dismiss such an idea. It would appear far more likely to be part of a device that spun.

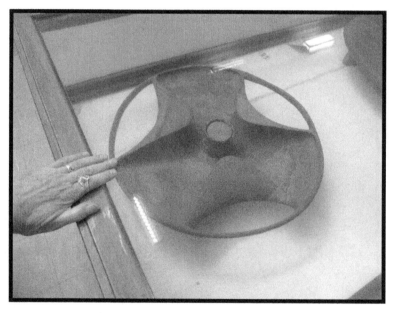

The schist disk with the author's hand to indicate size

In January 1936, this strange disk was unearthed at the plateau edge of North Saqqara, approximately 1.7 kilometers north of Djoser's Step Pyramid in Egypt. The discovery of the mysterious prehistoric artifact, that many considered a device of some kind, was made in the so-called Mastaba of Sabu (Tomb 3111, c. 3100-3000 BC) by a famous British Egyptologist Walter Bryan Emery (1902-1971). Sabu was the son of Pharaoh Aneddzhiba (5th ruler of the 1st dynasty of ancient Egypt) and a high official or administrator of a town or province possibly called 'Star of the family of Horus.' (38) The burial chamber had no stairway and its superstructure was completely filled with sand and stone vessels, flint knives, arrows, a few copper tools, and the most interesting schist bowl in fragments. The stone vessels numbered more than 10,000, some say in fact 40,000, and many appear to have turned on a lathe. These we will get to later.

The unearthed device named the schist disk is approximately 61 cm in diameter (24 inches), 1 cm thick (0.4 inches), and 10.6 cm (4.2 inches) in the center. It was manufactured by unknown means from supposedly a very fragile and delicate material requiring very tedious carving the production that would confound many craftsmen even today.

Photo of the disk showing the curvature of the blades

Our resident geologist Suzan Moore believes that it is not schist at all, but possibly very fine slate, or another type of metamorphosed sand or clay. Scientists do not think the object is a wheel, because the wheel appeared in Egypt about 1500 BC, during the 18th Dynasty. Egyptologist Cyril Aldred reached the conclusion that, independent of what the object was used for or what it represented, its design was without a doubt a copy of a previous, much older metallic object. Why did the ancient Egyptians bother to design an object with such a complex structure more than 5,000 years ago?

Photo of the smooth curvature of the bottom

Because the disk was found broken into several pieces, with some of the parts missing, it was reconstructed, likely by staff of the museum. Their repair job was quite poor, and thus accounts for why many observers think it was not originally symmetrical. The author does not agree.

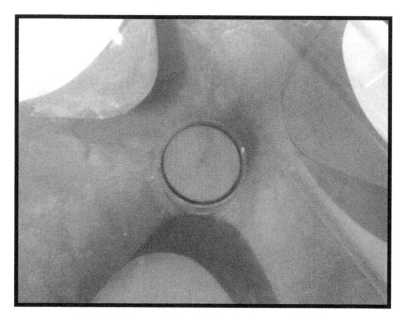

Detailed view of the blades and central hub

So, if we assume that this was not created as a lotus vase or incense burner and was in fact meant to rotate around the central hub, which has a wooden plug in it for some reason, what would it do? Some replicas have been made out of plastic or fiberglass, and attached to power drills or motors to make them spin. Since the blades on the disk are not angled in any way, it neither pushes nor pulls either air or water. No reasonable explanation so far has been given for what it did when rotating, though some speculate it created some sort of antigravity field that could have been used to move heavy objects.

Another curious artifact that looks somewhat like the above disk is believed to have been from the third dynasty and again appears to have been made from a fine and tightly compacted stone.

Curious small plate found in the Cairo Museum

This interesting little artifact, which is in a room adjacent to the above disk, appears to have been partially made on a lathe, as evidence by the circle in the center, but also shaped either by hand or some unknown technology. As you can see, it has five folds in the rim. It was clearly found broken as was the disk, but whether it was found in the same location or not is unknown by the author.

Also on display at the museum are several small, medium, and large plates, bowls, and other vessels which were found in the large cache at the massive ancient burial site called Saqqara, all seemingly piled together, some broken and others intact.

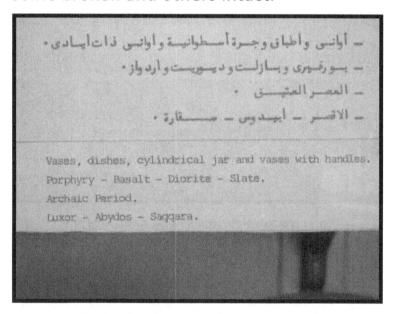

Label on display showing some of the stone types involved

There are some indications that they are from the very early dynastic period, or from an even earlier time, called the 'archaic period.' They appear to have in large part been made on some type of lathe. It is amazing that they are on display at all, because if machined, they do not fit in with the standard history of Egypt in any way, shape or form.

A sample of some of the stone vessels on display

Often 'out of place' artifacts such as these are never put on public display, because they do not fit the official paradigm. Many could only have been made using a lathe of high precision, not a simple potter's wheel, and could not have been made at the time stated. As many of these vessels are made of very hard stone, such as granite, diorite, and basalt, they could only be shaped with a tool made of, or incorporating, a harder material. As granite and diorite contain quartz, which has a hardness of 7 on the Mohs scale (10 being the hardest natural material known – diamond), only tools with corundum or diamond in their composition could be used to cut and shape them.

Assortment of jars and bowls on display

The previous photo shows jars and bowls made of a granite related stone, possibly diorite. The early dynastic Egyptians had at best quite crude bronze tools, and thus there is no way that they could have made these or many of the other vessels found in the Saqqara cache. Iron metal is singularly scarce in collections of Egyptian antiquities. Bronze remained the primary material there until the conquest by Assyria in 673-663 BC. The explanation of this would seem to lie in the fact that the relics are in most cases the paraphernalia of tombs, with funeral vessels and vases. Iron being considered an impure metal by the ancient Egyptians, it was never used in the manufacture of these or for any religious purposes. It was attributed to Seth, the spirit of evil who according to Egyptian tradition governed the central deserts of Africa. (39)

There are some claims that meteoric iron was used in ancient Egypt, perhaps as far back as 5000 years, but only for ornamental beads.

Finely shaped vase that clearly was made on a lathe

As we see in the photo above and next page, there is a very even crease near the lip of the vase, so finely done that likely it was done with a fine diamond-tipped tool.

The diamond core drill was invented and put to practical use in 1863 by Rodolphe Leschot, a French engineer. He used it for drilling blast holes for tunneling Mount

The previous photo shows jars and bowls made of a granite related stone, possibly diorite. The early dynastic Egyptians had at best quite crude bronze tools, and thus there is no way that they could have made these or many of the other vessels found in the Saqqara cache. Iron metal is singularly scarce in collections of Egyptian antiquities. Bronze remained the primary material there until the conquest by Assyria in 673-663 BC. The explanation of this would seem to lie in the fact that the relics are in most cases the paraphernalia of tombs, with funeral vessels and vases. Iron being considered an impure metal by the ancient Egyptians, it was never used in the manufacture of these or for any religious purposes. It was attributed to Seth, the spirit of evil who according to Egyptian tradition governed the central deserts of Africa. (39)

There are some claims that meteoric iron was used in ancient Egypt, perhaps as far back as 5000 years, but only for ornamental beads.

Finely shaped vase that clearly was made on a lathe

As we see in the photo above and next page, there is a very even crease near the lip of the vase, so finely done that likely it was done with a fine diamond-tipped tool.

The diamond core drill was invented and put to practical use in 1863 by Rodolphe Leschot, a French engineer. He used it for drilling blast holes for tunneling Mount

Cenis on the France-Italy border. Leschot patented the device in the United States in 1863 and it was reissued in 1869. Diamond tools for lathes were not invented until after this.

Finely shaped bowl with a very thin lip

The next photo not only shows more of the amazing vessels, but one in the center that has two drill holes in it. Clearly, we are looking at technology that existed before the time of the dynastic Egyptians, and only resurfaced in the late 19th century at the earliest. It is possible that these vessels were found by the early dynastic Egyptians and were then stored away at Saqqara until found by Emory. Since the majority of them were found broken, and again are very hard stone in many cases, it is possible that they were destroyed by some catastrophe in the very distant past.

More stone vessels showing, in one, drill holes

Another curiosity as pointed out by author Stephen Mehler, and first shown to him by his teacher Abd'El Hakim Awyan, are the presence of huge sarcophagi, made of wood, in the Cairo Museum. Though some academics insist that they were made for symbolic reasons, is it not possible that they once contained actual giant people?

Huge sarcophagi in the Cairo Museum

There are references in the oral tradition of very tall humans once living in Egypt, and many of you readers will have heard of such stories in North America, Europe, etc. It is too easy to dismiss such stories as being purely myth, especially when one sees the giant coffins in the Cairo Museum, as well as the many huge stone boxes found in other locations in Egypt.

Obvious drill holes in a large granite box

Finally, the photo above shows you several drill holes in the lip of a very large granite box. As has been stated earlier, the tool must be harder than the material being cut, shaped, or drilled.

We now move on to the most famous area of ancient Egypt, the Giza Plateau. Here we will find obvious examples of lost ancient high technology and feats that the dynastic Egyptians simply could not have achieved.

According to the standard story, the First Dynasty of ancient Egypt (or Dynasty I) covers the first series of Egyptian kings to rule over a unified Egypt. It immediately follows the unification of Upper and Lower Egypt, possibly by Narmer, and marks the beginning of the Early Dynastic Period, a time at which power was centered at Thinis.

The date of this period is subject to scholarly debate about the Egyptian chronology. It falls within the early Bronze Age and is variously estimated to have begun anywhere between the 34th and the 30th centuries BC. In a 2013 study based on radiocarbon dates, the beginning of the First Dynasty, the accession of Hor-Aha, was placed close to 3100 BC (3218–3035, with 95% confidence). (40) However, some other researches have consulted ancient 'kings list(s)' which state, with names, that there were thousands of years of rulers before this.

Our Khemit School group on the Giza Plateau in April 2016

Of course, the most famous and grandest constructions on the Giza Plateau are the three pyramids. The Great Pyramid of Giza (also known as the Pyramid of Khufu or the Pyramid of Cheops) is believed to be the oldest and largest of the three pyramids in the Giza pyramid complex bordering what is now El Giza, Egypt. It is the oldest of the Seven Wonders of the Ancient World, and the only one to remain largely intact. Based on a mark in an interior chamber naming the work gang and a reference to fourth dynasty Egyptian Pharaoh Khufu,

Egyptologists believe that the pyramid was built as a tomb over a 10-20 year period concluding around 2560 BC.

Clear evidence that the casing stones locked into the pyramid - they are not adhered to the surface

As the above official information indicates, though does not directly acknowledge, the early dynastic Egyptians could not have built the great pyramids at Giza. First, they admit that at that time, the Egyptians were an early Bronze Age culture, which tells us that they did not have the tools sufficient to cut all of the stone involved. The date of construction is based on simple painted glyphs hardly sufficient as evidence. Many great books are written about the Giza complex, so I will simply deal with the signs of lost ancient technology we see, and possible evidence of ancient cataclysm.

These are some of the major statistics about the Great pyramid, from the webpage: http://www.ancient-code.com/25-facts-about-the-great-pyramid-of-giza/.

1. The pyramid is estimated to have around 2,300,000 stone blocks that weigh from 2 to 30 tons each and there are even some blocks that weigh over 50 tons. Some say the figure is 2,500,000 or more. Some claim that all of the stones were made from a type of geopolymer concrete, yet that in itself would be a more complex procedure than cutting out the blocks from the Giza Plateau. As each block is a different shape and size, you would need about 2.5 million different forms to make them. Nonsense.

2. The so-called Pyramid of Menkaure, the Pyramid of Khafre, and the Great Pyramid of Khufu are precisely aligned with the Constellation of Orion. Theory by Robert Bauval.

3. The base of the pyramid covers 55,000 m² (592,000 ft²) with each side greater than 20,000 m² (218,000 ft²) in area.

4. The interior temperature is supposedly constant and equals the average temperature of the Earth, 20 degrees Celsius (68 degrees Fahrenheit). The author has never experienced this. In fact, it seems much hotter the farther up you go.

5. The outer mantle was composed of 144,000 casing stones, all of them highly polished and flat to an accuracy of 1/100th of an inch, about 100 inches thick and weighing approximately 15 tons each. The casing stone material is Tura limestone from a still active quarry in Cairo, not Giza.

6. The cornerstone foundations of the pyramid have ball and socket construction capable of dealing with heat expansion and earthquakes. This the author has not directly seen.

7. The mortar used between many of the core blocks is of an unknown origin. It has been analyzed, and its chemical composition is known, but it cannot be reproduced. It is stronger than the stone and still holding up today.

8. It was originally covered with casing stones made of highly polished limestone. These casing stones reflected the sun's light and made the pyramid shine like a jewel. They are no longer present, having been requisitioned to build mosques after an earthquake in the 14th century loosened many of them. It has been calculated that the original pyramid with its casing stones would act like gigantic mirrors and reflect light so powerful that it would be visible from the moon as a shining star on Earth. Appropriately, the ancient

Egyptians called the Great Pyramid 'Ikhet,' meaning the 'Glorious Light.' How the casing stones may have fallen off will be discussed later.

9. Aligned True North: The Great Pyramid is the most accurately aligned structure in existence and faces true north with only 3/60th of a degree of error. The position of the North Pole moves over time and the pyramid was exactly aligned at one time.

10. Center of Land Mass: The Great Pyramid is located at the center of the landmass of the Earth. The east/west parallel that crosses the most land and the north/south meridian that crosses the most land intersect in only two places on the Earth: one in the ocean, and the other at the Great Pyramid.

11. The four faces of the pyramid are slightly concave, or indented, with an extraordinary degree of precision. This is the only pyramid to have been built this way. This effect is not visible from the

ground or from a distance, but is from the air, and then only under the proper lighting conditions – at dawn and sunset on the spring and autumn equinoxes, and when the sun casts shadows on the pyramid.

Pyramid from above showing concave, or indented, sides

12. The granite box in the 'King's Chamber' is too big to fit through the passages and so it must have been put in place during construction.

13. The box was made out of a block of solid granite. This would have required, according to most academics, bronze saws 8-9 feet long set with teeth of sapphires. Hollowing out of the interior

would require tubular drills of the same material applied with a tremendous vertical force. As the early dynastic Egyptians were in a primitive stage of bronze development, they could hardly have had the ability to embed sapphires or diamonds into said material. As we have already seen, the first diamond tipped tools were not invented until well into the 19th century AD.

14. Microscopic analysis of the coffer reveals that it was made with a fixed-point drill that used hard jewel bits and a drilling force of two tons. Again, not something that the dynastic people of the period could have done.

15. The Great Pyramid supposedly had a swivel door entrance at one time. Swivel doors were found in only two other pyramids: Khufu's father and grandfather, Sneferu and Huni, respectively. The author never saw evidence of this, but an intriguing thought.

16. It is reported that when the pyramid was first broken into the swivel door, weighing some 20 tons, was so well balanced that it could be opened by pushing out from the inside with only minimal force, but when closed, was so perfect a fit that it could scarcely be detected and there was not enough crack or crevice around the edges to gain a grasp from the outside.

17. With the mantle in place, the Great Pyramid could be seen from the mountains in Israel and probably the moon as well.

18. The weight of the pyramid is estimated at 5,955,000 tons. Multiplied by 10^8 gives a reasonable estimate of the Earth's mass.

19. The Descending Passage pointed to the pole star Alpha Draconis, circa 2170-2144 BC. This was the North Star at that point in time. No other star has aligned with the passage since then. Could be a coincidence.

20. The southern shaft in the King's Chamber pointed to the star Al Nitak (Zeta Orionis) in the constellation Orion, circa 2450 BC. The Orion constellation was associated with the Egyptian God Osiris. No other star aligned with this shaft during that time in history. Again, could be coincidence.

21. Sun's Radius: Twice the perimeter of the bottom of the granite box times 10^8 is the sun's mean radius (270.45378502 Pyramid Inches* 10^8 = 427,316 miles).

22. The curvature designed into the faces of the pyramid exactly matches the radius of the Earth.

23. Khufu's pyramid, known as the Great Pyramid of Giza, is believed to be the oldest and largest, rising at 481 feet (146 meters). Archeologists say it was the tallest structure in the world for about 3, 800 years.

24. The relationship between Pi (p) and Phi (F) is expressed in the fundamental proportions of the Great Pyramid.

What these points indicate is that the Great Pyramid, as well as the other two large ones on the plateau, as regards scale, design, complexity, and composition, are among the greatest achievements in all of human history. The idea that a civilization in its early stages, experimenting with how to make bronze and with little knowledge of engineering could have achieved this no matter how many people they had working on it, must be thrown out - as well as the idea that they were constructed as tombs.

As the following photo shows, the casing stone, little of which remains intact, was not adhered to the surface, but interlocked in a three dimensional manner. This would have been a huge undertaking. Building the pyramid in 20 years, which is what most academics believe, would involve installing approximately 800 tons of stone every day. Since it consists of an estimated 2.3 million blocks, completing the building in 20 years would also involve moving an average of more than 12 blocks into place each hour, day and night. That means cutting them from the bedrock, moving them to the site, shaping them, and putting them into place with a high degree of accuracy.

Remaining casing stone locking into the Great Pyramid

Other estimates are that every two minutes a limestone block would have to have been cut, moved, dressed, and then accurately put into position.

Hard to see, but the remaining casing stones are within 1 degree of perfect level

Not only did the casing stones interlock with the main structure, but the stones of the core itself locked into the bedrock in a three dimensional manner. Such a feat is again complex, and could have been done for structural, or perhaps vibrational reasons, which we will get into.

The pyramid blocks interlocked with the bedrock

The Second Pyramid had at least one row of casing stones and perhaps more of granite. Whether this was an aesthetic or functional choice is unknown. Popular belief is that the granite all originated at the massive quarries at Aswan, but our geologist questions that, believing that at least some may have come from the eastern desert areas, perhaps the Sinai. What remains of the casing stone still fits so tightly that a human hair does not fit in the joints, and as has already been shown, the early dynastic people had no tools for such precise work.

Granite casing stone of the Second Pyramid

One curious thing about the Second Pyramid is that the white Tura limestone casing stones are still somewhat intact on the top, whereas the Great pyramid's surface has been largely stripped clean. Although the common belief is that the casing of all three pyramids was stripped by people, likely after the time of the dynastic

Egyptians, it could be that a cataclysmic earthquake, part of the cataclysm of about 12,000 years ago that we have been discussing, could have been why the casing stone came out of place.

The author in front of the Second Pyramid

With the pyramids essentially locked into the bedrock, a major earthquake would have caused the structures to shake very violently. This could have caused the casing stone joints to weaken and fall or even pop out. Far easier for those who utilized the material for the construction of Cairo to use what was on the ground, rather than trying to pry them out.

The Third pyramid also has some of its casing stone in place, but is very different in texture than the other two. Rather than being smooth, the surfaces of the casing, which is pinkish granite and presumably from Aswan, or the eastern desert, has a rippled finish, quite similar to what we see if the Inca Roca wall in Cusco Peru. Also, there are knobs, which the Inca Roca wall also has.

Attribution of the name Khafre is based on this one glyph

The author pointing out stones with knobs on them

Yousef Awyan pointing out the scoop marks

How this shaping was done is unknown, both in Peru or Egypt. The surface of the stone appears to also be much more eroded, and prone to erosion than that of the other two pyramids for some reason.

What exists above ground is impressive, of course, but most people do not realize that there is a massive system of shafts and tunnels below the plateau. Few of these entrances are in the area that the public gets to see, such as in the photo above, but there are far more impressive ones in other areas.

One of several entrances to the tunnel system beneath the Giza Plateau

I was first allowed to enter into this system in 2014, and in fact went three levels down. I was told by my guide, Yousef, that there were many levels below that. There appear to be more shafts in the area in between the Sphinx and Great Pyramid, along the ancient causeway which leads to the pyramid itself. Rather than going straight,

Several small shafts that lead to others and the tunnel system

perpendicular to one of the sides of the Great pyramid, the causeway goes at an angle to miss the Sphinx. This tells us that the Sphinx is older than the pyramids, which we will get into.

Shaft that leads to several side chambers, with the Sphinx head in the background

It is believed by some, and especially those of the Khemit School, that the function of the tunnels was to conduct water from the Nile. The shafts could have been access points for these tunnels, for maintenance perhaps. Most have grates on them supposedly for safety reasons, but they also prevent anyone from officially exploring them.

Massive shaft partly filled with sand

The theory goes that the tunnel system was created first, in the limestone bedrock, and conducted water under the Giza Plateau. The dimensions of the tunnels were specific, and set up vibrations that were then enhanced by the chambers and shapes of the pyramids.

Looking down into one of the deep shafts

What the vibrational energy or sound was used for is unclear. Perhaps to set up an energetic field that allowed the builders and other residents to access higher levels of consciousness. The British-born master machinist Chris Dunn, author of the Giza Power Plant, believes electricity may have been the function. Clearly, more research has to be done, but what is also inferred is that the original pyramids, as in those from

Abu Rawash to Dashur, were each specifically designed for a particular vibration. These were the original pyramids, that the dynastic people discovered when they first entered this area of the Nile some 5000 years ago. Over the course of time, they built their own, often made of mud brick, which have weathered far quicker than the originals, as they were very inferior. Some of the smaller chambers on the Giza Plateau were used by them as tombs - they simply found a use for what already was there.

It is believed that this energetic system was functioning fully until a great cataclysm. This could have been when portions of the polar ice packs melted, or perhaps were even vaporized due to solar plasma, according to geologists including Robert Schoch who, in a private conversation with the author, stated that this could have occurred in as little as two weeks – and the world changed.

Entrance to another massive shaft

Such a rapid alteration in the distribution of ice becoming water would have increased pressure on the Earth's crust, causing volcanoes to erupt, and the planet itself may have been thrown out of balance. It makes sense that originally it was perfectly vertical, with this imbalance of ice becoming water resulting in our present 23.5 tilt.

View of the causeway leading to the Sphinx

This would explain, for example, why woolly mammoths were found in the permafrost seemingly flash frozen with buttercups in their mouths. They went suddenly from a temperate climate to the frozen north. Also, this could account for tales of such ancient civilizations like Atlantis, inundated by water over a very short period of time.

With the planet suddenly shifting its axis, the energy system of the Giza Plateau would immediately shut down, being rendered useless, and its residents either perished, or were forced to move.

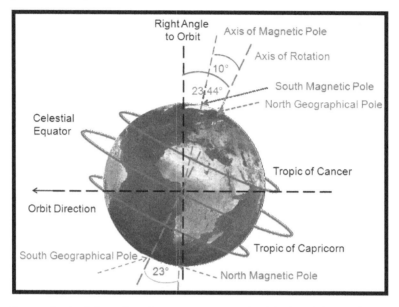

Seems logical that the planet was created with the right angle to orbit

There are some profoundly obvious examples of lost ancient high technology scattered around the pyramids. As the next photo shows, saws on a massive scale were clearly in the area at the time, or perhaps even before the construction of the pyramids.

Obvious saw cut in basalt stone near the Great Pyramid

Though some people claim these cuts to be recent, such ideas are not backed by any evidence whatsoever. Next to the Great Pyramid there are the remains of a large basalt platform which has been plundered over the course of possibly millennia for building stone in other areas of Giza, or possibly Cairo. Here we find several massive saw cuts, by both straight saws and circular ones, whose cutting rate is greater than any such 21st century tool - as much as 2 mm per revolution.

Obvious evidence of an ancient saw at work

There are also numerous ancient drill holes seemingly with a penetration rate of 2 mm per revolution. As well, the large stone box inside the so-called 'King's Chamber' in the Great Pyramid has saw marks in it that would appear to be original.

Other factors that indicate that the pyramids of the Giza plateau bear no relationship to the dynastic works are clear and numerous. The most obvious is that there is no ornamentation inside any of these three structures. The dynastic people literally put glyphs anywhere they would fit; on walls, columns, sculpture, etc. Also, the pyramids were the greatest undertaking in the entire area, and they were made first?

Further, the simple geometry of the pyramids and their enormity are very unlike other major sites in Egypt. As we travel into Upper Egypt, against the flow of the Nile, there are anomalous artifacts, such as massive obelisks, which are clearly pre-dynastic and could very well be contemporary with the ancient pyramids.

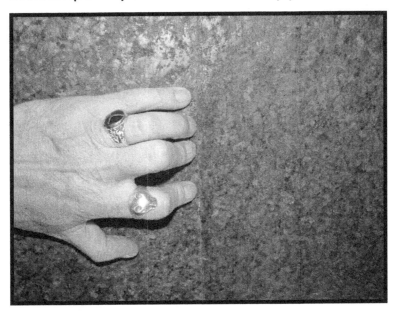

Wall inside the King's Changer - amazing precision

All alone in the glyph-less grand gallery of the Great Pyramid

According to oral tradition, and communicated to the author by fellow author Stephen Mehler, the pyramids in the Abu Rawash to Dashur area are in the oldest part of Egypt. This area was originally known as Bu Wizzer, translated as meaning the 'Land Of Osiris.' It is from Wizzer that we get the English words 'wizard' and 'wisdom.' The ancient pre-cataclysmic works are fewer the farther you move away from Bu Wizzer, but are still present. It is a sad fact that most of Egypt's great obelisks were either destroyed, or carted off to foreign countries. Those that still exist are a sight to behold.

Tops of two obelisks at Karnak

As the classic obelisks were always made from one piece of stone, weighing hundreds of tons, it is not conceivable that they were of dynastic origin. One theory believes that they were part of the ancient pyramid energy system. If the pyramids emitted energy, especially vibration, then it is quite possible that the obelisks were made by the same builders as energy receivers, much like how television, radio, and Wi-Fi works today. We will get into this more later when we discuss Karnak in full.

Patricia Awyan of the Khemit School inside the Valley Temple

According to most academics, the Valley Temple was part of the funerary complex of Khafre including, along with the pyramid (with its' so-called 'burial chamber'), a mortuary temple (joining the pyramid on its east side), and a covered causeway leading to the Valley Temple itself. The purpose of these valley temples has been debated. They could have been used for the mummification process, or perhaps for the 'opening of the mouth' ceremony, when the 'ka' entered the deceased person's body. This temple is in an excellent state of preservation, having been buried by desert sand until the 19th century.

We have already fully explained that the dynastic Egyptians could not have constructed sites such as this, as they did not have the tools to quarry, shape, and move the granite it is composed of. The idea that massive pieces of stone such as this were floated down the Nile on barges, or dragged across the desert on sledges of some kind, is preposterous. For a start, where would the wood have come from?

Anyone who has visited Egypt will note that the only trees of any appreciable size are date palms, which would be unsuitable for boat construction because it is very soft and stringy in nature.

Engineer Alex measuring the flatness of a surface

Possibly the cedars of Lebanon? It is known that the Egyptians did use this wood for boat construction, but the size of a ship or barge needed to move one of the large obelisks would have been massive, and transporting the wood 600 kilometers from the forest would have been a major task.

Pharaoh Khafre's Valley Temple was built in the mid to late 26th century BC according to Egyptologists, which of course was at a stage where they only had bronze tools. The temple is constructed of a limestone core of huge blocks, many weighing between 100 to 150 tons. The blocks were quarried from the plateau surrounding the Great Sphinx that, along with its temple, lies adjacent to Khafre's valley temple. The floors throughout the temple are paved with travertine, which is a type of marble.

Exterior of the Valley Temple, once cased in granite blocks

Why the walls were made so thick is a quandary. Not only did you originally have precision shaped and fitted granite stones on the outside and inside, but also a core made of huge limestone blocks. The Valley Temple is next to the Sphinx Temple, the latter being right in front of the Sphinx itself. According to Yousef Awyan, and blatantly obvious when you see the site for yourself, they were in fact one complex.

Glimpse into the Sphinx Temple with the Valley Temple at right

The Sphinx Temple has not fared as well as the Valley Temple, most of its granite being stripped off, both outside and inside. The amount of weathering of the internal limestone, though a soft stone, is extreme to say the least, and looks like several thousands of years of exposure to the elements. An astonishing fact about the Valley Temple is that its construction appears to have been in two stages - the granite casing stones, on the side facing the limestone blocks, was cut in peaks and troughs to fit the existing erosional patterns of the limestone core blocks. In other words, when the granite facing was added, the limestone core was already heavily eroded, implying that the core blocks were already vastly old when this renovation took place.

Egyptologists, as mentioned earlier, attribute the building of the Valley Temple to the Pharaoh Khafre (2520-2494 BC). This is because a diorite statue of this pharaoh was found inside the temple, buried upside-down in a pit. As far as convincing evidence goes, this hardly ranks very highly – indeed, there exists a passage of text in hieroglyphs on an artifact known as the Inventory Stella, which implies that the Valley Temple (and the Sphinx) already existed during the reign of Khufu, who ruled more than 30 years before Khafre. (41)

View of the Sphinx Temple from the front - notice the extreme erosion

The Valley and Sphinx Temples are contemporary with the Sphinx, as they are comprised of massive stones removed from the Sphinx enclosure when the Sphinx was being shaped. Bronze tools can cut and shape limestone, but the sheer scale of the stones removed and their number would be a herculean task for any civilization.

It was the ground-breaking work of geologist Robert Schoch, with the heeding of John Anthony West, that turned the Egyptologists' world upside down. In his own words:

Khemit School with the Inventory Stelae

'In 1990 I first traveled to Egypt, with the sole purpose of examining the Great Sphinx from a geological perspective. I assumed that the Egyptologists were correct in their dating, but soon I discovered that the geological evidence was not compatible with what the Egyptologists were saying. On the body of the Sphinx, and on the walls of the Sphinx Enclosure (the pit or hollow remaining after the Sphinx's body was carved from the bedrock), I found heavy erosional features (seen in the accompanying photographs) that I concluded could only have been caused by rainfall and water runoff. The thing is, the Sphinx sits on the edge of the Sahara Desert and the region has been quite arid for the last 5000 years. Furthermore, various structures securely dated to the Old Kingdom show only erosion that was caused by wind and sand (very distinct from the water erosion). To make a long story short, I came to the conclusion that the oldest portions of the Great Sphinx, what I refer to as the core-body, must date back to an earlier period (at least 5000 BC, and maybe as early as 7000 or 9000 BC), a time when the climate was very different and included more rain.

"Many people have said to me that the Great Sphinx cannot be so old, in part because the head is clearly a dynastic

Egyptian head and the dynastic period did not start until about 3000 BC. In fact, if you look at the current Great Sphinx you may notice that the head is actually too small for the body. It is clear to me that the current head is not the original head. The original head would have become severely weathered and eroded. It was later re-carved, during dynastic times, and in the re-carving, it naturally became smaller. Thus, the head of the Great Sphinx is not the original head. In fact, the Sphinx may not have originally been a sphinx at all. Perhaps it was a male lion." (42)

Others, notably Stephen Mehler, and based on the teachings of Abd'El Hakim Awyan, believe the Sphinx was carved in three phases, and that it was originally a female lion, the personification of Tefnut. According to the indigenous tradition, Nut is the sky or cosmos, representing all that is unmanifested. When Nut (pronounced 'not') wanted to appear in a physical form, she spat upon the Earth and where this was became the limestone outcropping that eventually was carved into the form of the Sphinx.

The Sphinx Temple blocks, put into place as cut from the enclosure

The Sphinx showing extreme weathering and reconstruction

Signs of extreme weathering in the Sphinx enclosure

Weathering of the body shows that the head was recarved

Horizontal weathering by sand and wind, and vertical by rain

Dr. Schoch continues:

"To further test the theory of an older Sphinx, we carried out seismic studies around the base of the statue to measure the depth of subsurface weathering. Basically, we used a sledgehammer on a steel plate to generate sound waves that penetrated the rock, reflected, and returned to the surface. This gave us information about the subsurface qualities of the limestone bedrock. When I analyzed the data, I found that the extraordinary depth of subsurface weathering supported my conclusion that the core-body of the Sphinx must date back to 5000 BC or earlier.

"During the seismic studies we also discovered clear evidence of a cavity or chamber under the left paw of the Sphinx. For what it is worth, some have suggested to me that this may be a 'Hall of Records' (at the time I was not aware of Edgar Cayce's predictions along these lines). Additionally, we found some lesser (and previously known) cavities under and around the Sphinx, and the data also indicates that there may be a tunnel-like feature running the length of the body.

"Back in the early 1990s, when I first suggested that the Great Sphinx was much older than generally believed at the time, I was challenged by Egyptologists who asked, 'Where is the evidence of that earlier civilization that could have built the Sphinx?'

Boardwalk in place to hide the entrance to chambers and tunnels below

"They were sure that sophisticated culture, what we call civilization, did not exist prior to about 3000 or 4000 BC. Now, however, there is clear evidence of high culture dating back over 10,000 years ago, at a site in Turkey known as Göbekli Tepe. A major mystery has been why these early glimmerings of civilization and high culture disappeared, only to reemerge thousands of years later."

Evidence? How about the 10,000+ turned vessels from Saqqara

By asking for "evidence of that earlier civilization," academics are ignoring such things as the thousands of turned hard stone bowls from Saqqara that are in the Cairo Museum, as well as the obvious megalithic evidence we have been discussing. Rarely if ever have they seemed to have consulted with a master machinist to ask how complex a lathe would have to be to create the stone vessels as seen above and earlier, or an architect to see how complex the drawings would have to be for the Great pyramid, or an engineer to find out how multi-ton blocks could have been transported from the Sinai Desert, or a stone mason to see if such tasks could be recreated today. Giza by itself could be the subject of volumes of discussion, but let us move on to Dashur.

It is claimed by Egyptologists that the Bent Pyramid was built under the Old Kingdom Pharaoh Sneferu (c. 2600 BC). A unique example of pyramid development in Egypt, this was the Second Pyramid built by Sneferu, they claim. The Bent Pyramid rises from the desert at a 54-degree inclination, but the top section is built at the shallower angle of 43 degrees, lending the pyramid its very obvious 'bent' appearance.

The Bent Pyramid at Dashur

Archaeologists believe that the Bent Pyramid represents a transitional form between step-sided pyramids, like one at Saqqara, and smooth-sided pyramids, like those of the Giza Plateau. It has been suggested that due to the steepness of the original angle of inclination the structure may have begun to show signs of instability during construction, forcing the builders to adopt a shallower angle to avert the structure's collapse. This theory appears to be borne out by the fact that the adjacent Red Pyramid, built immediately afterwards by the same Pharaoh, was constructed at an angle of 43 degrees from its base.

All four corners show extreme damage to the casing stones

There appears to be no evidence whatsoever that the Bent and Red Pyramids were constructed during the time of Sneferu, and in fact where would the skilled workforce have come from? Nearby there is also what is often called the Black Pyramid. The Black Pyramid was built by King Amenemhat III during the Middle Kingdom of Egypt (2055-1650 BC). It is one of the five remaining pyramids of the original eleven pyramids at Dahshur. Originally named Amenemhet is Mighty, the pyramid earned the name Black Pyramid for its dark, decaying appearance as a rubble mound. It is possible that this pyramid was built on top of older foundations as it has complex passageways underground. The fact that its exterior was largely mud brick, has decayed badly, and was built after the Red and Bent ones, even according to standard Egyptology, shows that the builders' techniques had also decayed.

Photo of the tightness of the casing stone fit

The construction techniques and precision used to make the Bent and Red Pyramids are the same as those of the Giza Plateau, so it is likely that all five are contemporary with each other. And, they all are in the zone described earlier as Bu Wizzer, which according to author Stephen Mehler, is the oldest area of occupation in Egypt.

Rather than being a failed experiment by Sneferu's architects and engineers, the pyramids at Dashur are likely pre-dynastic and were part of the massive energetic system that included the Giza Plateau and the farther pyramid at Abu Rawash. According to Stephen Mehler, the term 'Sneferu' translates into English as 'double harmony,' and could thus describe the fact that perhaps rather than resonating to a specific frequency, such as the Great Pyramid, it resonated to two.

Photo of damage to one of the four corners of the Bent Pyramid

The fact that all four corners show somewhat even damage is curious. Though most academics attribute this stone removal to later people harvesting the stone to build other buildings, engineers have told the author that a violent shake, such as that from a cataclysmic earthquake, could have blown the four corners out. It does not seem logical that people trying to take the structure apart would work on more than one corner at a time.

Partial view of the Red Pyramid

Named for the rusty reddish hue of its red limestone stones, the Red Pyramid is the third largest Egyptian pyramid, after the Great and Second ones at Giza. It was not always red. It used to be cased with white Tura limestone, but only a few of these stones now remain at the pyramid's base, at the corner. During the Middle Ages it is believed that much of the white Tura limestone was taken for buildings in Cairo, revealing the red limestone beneath. Egyptologists disagree on the length of time it took to construct. Based on quarry marks found at various phases of construction, Rainer Stadelmann estimates the time of completion to be approximately 17 years while John Romer, based on this same graffiti, suggests it took only ten years and seven months to build. (43)

Appearance of extreme erosion on the Red Pyramid

What the author finds odd about the exterior of the Red Pyramid is that there are basically no casing stones at all to be found on the site. Any remnants of the white Tura limestone are basically the size of chips that you can hold in your hand. Why the Red Pyramid's casing was completely harvested, while only a small percentage of the nearby Bent Pyramid's exterior is gone, is a mystery.

More casing stones missing from the upper half of the Bent Pyramid

The above photo shows you one side of the Bent Pyramid where the casing stones are missing from the top down to where the angles change. This would not be logical way for people to harvest the skin of the pyramid. An alternative theory could be that a massive shock, such as a catastrophic earthquake, was more effective at disturbing the shallower angle than the steeper one, and that is why we see the upper half of the Bent Pyramid, on one side at least, and the entire Red Pyramid without casing stones.

The Bent Pyramid is off limits as regards entering by the public, but the Red Pyramid's core is open.

The author about to explore the core of the Red Pyramid

As can be seen in the previous photo, the shaft that down into the Red Pyramid is reasonably large. It then connects with three internal chambers, which have amazing sound resonance. Each time we visit, local expert Yousef Awyan escorts us. He knows the exact pitch that makes these chambers and those on the Giza Plateau, resonate the strongest. The idea that such a sonic effect was not a planned aspect of the chambers is ludicrous.

The walls of each chamber taper as they go up. Some would say this was done to strengthen the chamber from possible collapse, but it was more likely done to enhance the sound.

We now head to the massive ancient site of Saqqara, which is in fact one of the largest ancient graveyards on the planet.

The Pyramid of Djoser (or Zoser), or the Step Pyramid, is an archeological remains in the Saqqara necropolis, and likely the most famous. It was supposedly built during the 27th century BC for the burial of Pharaoh

Djoser by Imhotep, his vizier, and is the central feature of a vast mortuary complex in an enormous courtyard surrounded by ceremonial structures and decoration.

Inside the first chamber - note the upper tapering of the walls

This first Egyptian pyramid of the dynastic people consisted of six mastabas (of decreasing size) built atop one another in what were clearly revisions and developments of the original plan.

Patricia Awyan and the Khemit School approaching the Djoser Pyramid

The pyramid originally stood 62 meters (203 feet) tall, with a base of 109 × 125 m (358 × 410 feet), and was clad in polished white limestone. The step pyramid (or proto-pyramid) is considered the earliest large-scale cut stone construction, although the pyramids at Caral in South America are contemporary. Those at Bu Wizzer are of course older, and Djoser's architects likely tried to replicate their shape, with very limited success.

Broken fragments of the turned vessels discussed earlier

It is in this general area that the 10,000 to 30,000 or more stone vessels were found, discussed earlier. As the photo above shows, many pieces of once highly polished and clearly lathe-turned bowls, plates, and vases litter the area, and are stacked in large piles. Though the Djoser Pyramid is somewhat interesting, it is clearly from the dynastic period and thus not pre-cataclysmic. What many people have not seen, since it has been closed until recently, is the Serapeum - a tunnel system underground at Saqqara.

The author next to one of the giant boxes in a tunnel of the Serapeum

Egyptologists claim that it was the burial place of the Apis bulls, which were incarnations of the deity Ptah. It was believed that the bulls became immortal after death as Osiris Apis, shortened to Serapis in the Hellenic period. The most ancient burials found at this site date back to the reign of Amenhotep III.

In the 13th century BC, Khaemweset, son of Ramesses II, ordered that a tunnel be excavated through one of the mountains, with side chambers designed to contain large granite sarcophagi weighing up to 70 tons each, not including their lids, which supposedly held the mummified remains of the bulls. (44)

A second tunnel, approximately 350 m in length, 5 m tall and 3 m wide (1,148.3 × 16.4 × 9.8 feet), was supposedly excavated under Psamtik I, and later used by the Ptolemaic Dynasty.

One of the great boxes, with lid, in an alcove of the Serapeum

The temple was discovered by Auguste Mariette, who had gone to Egypt to collect Coptic manuscripts but later grew interested in the remains of the Saqqara necropolis. In 1850, Mariette found the head of one sphinx sticking out of the shifting desert sand dunes, cleared the sand, and followed the boulevard to the site.

After using explosives to clear rocks blocking the entrance to the catacomb, he excavated most of the complex. Unfortunately, his notes of the excavation were lost, which has complicated the use of these burials in establishing Egyptian chronology. Mariette supposedly found one undisturbed and sealed box, the contents of which are now said to be at the Agricultural Museum in Cairo. However, when Khemit School member Mohamed Ibrahim went to that museum to inquire about the box's contents, the staff had no clue what he was talking about. Thus, there is no evidence that they ever existed.

Khemit School members inspecting one of the massive boxes

As Mariette found all but one of the boxes open, or with lids sliding ajar, and used explosives to blow open the one that was sealed, there is no evidence that they in fact ever contained anything. The idea that they had been previously looted is sheer speculation, and the opening gap between the lids and boxes is barely large enough for a normal sized human to crawl through.

The boxes appear to weigh approximately 70 tons, with the lid adding a further 30 tons. Each lid was cut from the box itself, as was also evidenced in the box with the broken lid located inside the Cairo Museum. In modern times, heavy equipment was supposedly brought into the Serapeum to move some of the lids, but to no avail. So how could ancient tomb robbers have supposedly shifted them?

The author on one of the great boxes - note the orbs or dust in the air

All but one of the more than 20 boxes was made from granite, grano-diorite, or diorite, and our resident geologist Suzan Moore does not believe that they all came from the great quarries of Aswan, 500 miles to the south. Some may have come from the eastern desert - but how? Once again, iron and hardened steel were not present in any quantity in the 13th century BC when these boxes were supposedly made, and no such tools of these materials have ever been put on public display.

Author Stephen Mehler under the polished lid of one box

The arid environment of this part of Egypt surely would have reduced the possibility that such tools would have rusted away to nothing.

Another one of the massive boxes

The boxes are in different stages of completion. A few have been precision finished with a glossy sheen to the surface, while others still have a rough feel to them, and some do not have lids whatsoever. It would appear that at least the finishing work was done where the boxes are located, but as these tunnels are underground, and if the lights are turned off become pitch black, how could the workers see what they were doing?

Very tight and fine crystal structure - thus not typical Aswan granite

The ceilings of the tunnels have supposedly never been cleaned, and there is no appearance of any soot from torches that dynastic workers would have used, so the mystery of how the boxes were moved into the tunnels and their final resting places is still present. As well, as the next photo shows, the tunnels are barely large enough for the boxes to be moved in them.

Members of the Khemit School giving sense of scale

There are etchings, as in glyphs, on at least one of the Serapeum boxes, and deciphering these were one of the evidences used by Egyptologists to attempt to date them. However, as the next photo shows, the quality of the glyphs are far inferior to the precise surfaces, thus were not done by the same craftsmen. It is obvious to the author that the dynastic

Egyptians found the Serapeum boxes, and etched their crude glyphs onto them.

Very crudely etched glyphs and somewhat wandering lines

The finished boxes have a very fine finish inside, with no appearance of tool marks. Also, they greatly amplify sound if you hit the right resonance with your voice. The author felt almost deafened when he was able to achieve the proper pitch, and luckily caught all of this in video footage that is available at the YouTube channel: https://www.youtube.com/user/brienfoerster.

The corners of the interior of one of the boxes that the author inspected were not sharp, but slightly rounded, and by using fingertip touch as a gauge were very consistent, as if done with a router or other rotating tool. However, whatever technology was used did not leave any tool marks that the naked eye could see.

In one of the boxes - note the sheen on the wall

The exterior surfaces in some cases are very curious. Rather than having the flat precision of the interior, there have polished depressions, which so far have not been explained. Whoever the dynastic artisans were that etched in the crude glyphs they simply carved away continuously whether they were working on a flat surface or curved.

Glyphs carved into the flat surfaces and depressions

Thanks to local connections, at the end of one of the tunnels, a locked steel door was opened and we were allowed to see that the tunnel continued on for another hundred or so feet, where it had been bricked up at the end. Supposedly beyond the bricks it simply keeps going.

Inside one of the unfinished boxes of the Serapeum

The author on top of one of the unfinished boxes

The author has not found anyone with a reasonable explanation as to what the boxes of the Serapeum were originally used for, but it does appear that they are pre-dynastic. They could perhaps be contemporary with the original Bu Wizzer pyramids.

Behind the locked door...

Now we move on to Upper Egypt, as in farther south, to explore evidence of pre-dynastic masterpieces and cataclysmic damage.

The Dendera Light is a supposedly ancient Egyptian electrical lighting technology depicted on three stone reliefs (one single and a double representation) and other surfaces in the Hathor Temple at the Dendera Temple complex located in Egypt.

One of the famous 'light bulbs' at Dendera

The sculpture became notable among 'alternative' historians because of the resemblance of the motifs to some modern electrical lighting systems. The view of Egyptologists is that the relief is a mythological depiction of a djed pillar and a lotus flower, spawning a snake within, representing aspects of Egyptian mythology. The djed pillar is a symbol of stability, which is also interpreted as the backbone of the God Osiris. In the carvings, the four horizontal lines forming the capital of the djed are supplemented by human arms stretching out, as if the djed were a backbone. The arms hold up the snake within the lotus flower. The snakes coming from the lotus symbolize fertility, linked to the annual Nile flood.

Another depiction of an ancient 'light bulb'

In contrast to mainstream interpretations, there is an alternative hypothesis according to which the reliefs depict Ancient Egyptian electrical technology, based on comparison to similar modern devices (such as Geissler tubes, Crookes tubes, and arc lamps). J. N. Lockyer's passing reference to a colleague's humorous suggestion that electric lamps would explain the absence of lampblack deposits in the tombs has sometimes been forwarded as an argument supporting this particular interpretation (another argument being made is the use of a system of reflective mirrors). What is actually being represented is unclear. However, there is also an ancient zodiac at Dendera.

Copy of the Dendera Zodiac - the original was taken by Napoleon

The sculptured Dendera Zodiac is a widely known Egyptian bas-relief from the ceiling of the pronaos (or portico) of a chapel dedicated to Osiris in the Hathor Temple at Dendera, containing images of Taurus (the bull) and the Libra (the scales). This chapel was begun in the late Ptolemaic period. The pronaos was added by the emperor Tiberius. This led Jean-François Champollion to date the relief correctly to the Greco-Roman period, but most of his contemporaries believed it to be of the New Kingdom. The relief, which John H. Rogers characterized as "the only complete map that we have of an ancient sky," has been conjectured to represent the basis on which later astronomy systems were based. (45) It is believed by some researchers that the Dendera Zodiac could be a copy of a much earlier one, and that astrology and astronomy first developed after the great cataclysm of about 12,000 years ago, due to the fear that such an event from the sky may repeat itself.

Somewhat near to Dendera is Abydos. The Abydos 'Helicopter' (a.k.a. Abydos Submarine, Abydos Jet Plane, Abydos UFO, etc.) is a pseudoscientific modern myth that has been spread rapidly via the internet concerning the singular appearance of a re-carved inscription in the mortuary temple of Seti I in Abydos, Egypt.

The famous and controversial glyphs at Abydos

It is claimed this carving depicts various high-tech or alien technologies, such as submarines, jet planes, and UFOs, thus playing into the theories that Ancient

Egyptian civilization was either influenced or founded by aliens or Atlanteans.

A wall section showing obvious evidence of recurving

The temple was built during the 19th Dynasty of Ancient Egypt, in the post-Amarna New Kingdom. It was built as a cenotaph and mortuary temple for Seti I, linking him to the cult of Osiris, which had a major presence at Abydos, and the temple features a prominent Osirian theme throughout, although other deities, as well as Seti I himself, were worshipped here. The temple was not completed in the lifetime of Seti I, but was completed by his son, Ramesses II, early in the reign of the latter. The work of Ramesses II was inferior to that of his father, and it is easy to tell, even without reading the temple inscriptions, in which reign each section was completed. As a result of this 'shoddy' work, some inscriptions were re-carved, hastily chiseled out, modified using plaster infill, or even just plastered over and new inscriptions chiseled into the plaster that, over millennia, crumbled or dried, falling away from the stonework. Of much greater interest to us is the Osirion, also found at Abydos.

The Osirion was supposedly an integral part of Seti I's funeral complex, and was also supposedly built to resemble an 18th Dynasty Valley of the Kings tomb. It was discovered by archaeologists Flinders Petrie and Margaret Murray, who were excavating the site in 1902 - 1903. The Osirion was originally built at a considerably lower level than the foundations of the temple of Seti, who ruled from 1294 - 1279 BC.

The author inside the Osirion, thanks to local connections

While there is disagreement as to its true age, Peter Brand says it "can be dated confidently to Seti's reign," despite the fact that it is situated at a lower depth than the structures nearby, that it features a very different architectural approach, and that it is frequently flooded with water, making carving it impossible had the water level been the same at the time of construction.

Another view of the interior of the Osirion

It is also extremely megalithic, and does not have all of the glyph ornamentation of the rest of the vast complex at Abydos. The vertical columns, weighing at least 60 tons each, are granite or possibly grano-diorite, and thus would have had to have been brought to this location from either Aswan or the eastern desert. Also, the walls are made from quartzite. The major ancient quarry of such stone being near Cairo, meaning that the granite was likely moved at least 400 kilometers and the quartzite about 200 kilometers. Most of the rest of the Abydos complex was either made of sandstone or limestone.

Depictions of the 'flower of life' painted on a column

Unfortunately, the most famous thing about the Osirion is the painting of the 'flower of life' symbol at different locations on the surfaces in red ochre paint. A number of these patterns can be seen on more than one of the columns of the Osirion, which is clearly ancient megalithic and pre-dynastic. However, the Greek text shows that this must have been placed here at a much later date. Possibly during the Ptolemaic period in Egyptian history, which lasted from 332 - 30 BC. It is just possible that this design could have been influenced by Pythagoras, which would push the dating back to around 547 BC. Claims that they were burned into the surface with a laser is simply not true.

Another view of the interior of the Osirion

The Osirion is very similar in basic construction to the Valley and Sphinx

Temples of the Giza Plateau, and are likely contemporary in age. There are no Egyptian glyphs on the surfaces of the columns in any of these three structures. The original function of the Osirion is unclear. It was seemingly built underground for a specific reason, but what that would be is unknown. Normally it has water in it, and there is what would appear to be an ancient tunnel or possibly even water channel that leads to its only entrance/exit, as seen in the next photo.

Just in front of the tunnel leading into the Osirion is another that is perpendicular to it. It is at an angle and may have originally led to the surface, but the actual entrance seems to be buried in the sand. Both tunnels are tall enough to allow a normal-sized person to walk through.

Tunnel leading to the Osirion

Perpendicular tunnel in front of that seen in the previous photo

On this last visit, I saw that the lintel of the entrance to the Osirion has a very long obvious saw cut on the backside of the quartzite stone. As quartzite has a hardness of 7 on the Moh's scale out of 10, there is no way that the dynastic Egyptians could have made this mark. It is the longest straight saw cut that the author has seen in Egypt so far.

Obvious saw cut near the top of the entrance lintel

Now we will move on to Karnak, where there appear to be obvious signs of possible cataclysmic damage to the core of this vast ancient site.

The Karnak Temple Complex, commonly known as just Karnak, comprises a vast mix of decayed temples, chapels, pylons, and other buildings. Building at the complex began during the reign of Senusret I in the Middle Kingdom, and continued into the Ptolemaic period, although most of the extant buildings date from the New

Kingdom. The area around Karnak was the Ancient Egyptian Ipet-isut ('The Most Selected of Places') and the main place of worship of the eighteenth dynasty Theban Triad with the God Amun as its head. It is part of the monumental city of Thebes. The Karnak complex gives its name to the nearby and partly surrounded - modern village of El-Karnak, 2.5 kilometres (1.6 miles) north of Luxor.

The history of the Karnak complex is largely the history of Thebes and its changing role in the culture. Religious centers varied by region, and with the establishment of the current capital of the unified culture, that changed several times. The city of Thebes does not appear to have been of great significance before the 11th Dynasty, and previous temple building here would have been relatively small, with shrines being dedicated to the early deities of Thebes - the Earth Goddesses' Mut and Montu. However, in the core area we find definite pre-dynastic examples of stone work.

Classic main entrance into Karnak

The largest example of a core drill so far

It is in the core that we find a path leading from east to west. This is through the main entrance and leads all the way to the back of the vast complex. We walk through many sandstone columns that were made by the dynastic Egyptians, but the core, sometimes referred to as the 'Holy of Holies' is of different construction.

Interior of the Holy of Holies composed of solid granite

Unlike the rest of Karnak, there is a lot of granite in the central core, and thus, would be the oldest part of the complex. Almost every stone structure and obelisk have severe damage - likely not the result of vandalism or weathering, but of a catastrophic force, perhaps heat.

Strange 'weathering patterns' on a block of black granite

We first were shown, by Yousef Awyan, an area to the left side of the Holy of Holies about three years ago that had large stones, perhaps part of an obelisk, with very strange signs of degredation. Both geologists Suzan Moore and Robert Schoch have been shown these, and initially had no logical explanation of what had happened to the stone. There are large cracks that would not have been the result of sun or rain, and the size of the crystals inside are much larger than on the surface. Some have surmised that this was the result of a sudden absorption of energy from an unknown source, but on our 2016 trip a more logical, and very controvercial theory came into the author's mind.

Photo clearly showing the internal crystals to be larger than the surface

It is Robert Schoch's theory that plasma ejections from the sun 11,700 years ago were responsible for the rapid ending of the last ice age, as well as raising the ocean level of the planet very dramatically. Plasma consists of electrically charged particles. Familiar plasma phenomena on Earth today include lightning and auroras, the northern and southern lights, and upper atmospheric phenomena known as sprites. In the past, much more powerful plasma events sometimes took place, due to solar outbursts and coronal mass ejections (CMEs) from the sun, or possibly emissions from other celestial objects. Powerful plasma phenomena could cause strong electrical discharges to hit Earth, burning and incinerating materials on our planet's surface.

Plasma hitting the surface of Earth could heat and fuse rock, incinerate flammable materials, melt ice caps, vaporize shallow bodies of water creating an extended deluge of rain, and send the climate into a warming spell. The release of pressure that follows the melting of thousands of meters thick ice sheets can induce earthquakes and even cause hot rock under pressure to melt and erupt to the surface as volcanoes. It is possible that the strange effects on the stone at the Holy of Holies, and other sites nearby, are evidence of Dr. Schoch's theory.

More examples of the strange decomposing stone

The plasma event of 9700 BC eradicated advanced civilizations and high cultures of the time, and the radiation emanating from the plasma may have affected mental and psychical abilities. This could be the basis for the nearly universal myth of a Golden Age, a time when beings on Earth had mental abilities far surpassing those of later times.

Extreme erosion on a vertical stone near the entrance to the Holy of Holies

The 9700 BC event may be the original basis for the Atlantis legends as the timeframe fits well with Plato's account.

Path to the Holy of Holies facing east

The plasma did not strike all of the planet, but some locations. It is possible that came from the east, because that is where the sun rises. At Karnak, the path to the Holy of Holies is about 20 degrees off of true east/west. If the planet was vertical in terms of its axis before the cataclysm of 11,700 years ago, and is now 23.5 degrees off, it could make sense that prior to the cataclysm, the path was indeed true east/west. The plasma would scorch and damage everything in its path, and vaporize any forms of life nearby.

Damaged vertical stones on the left and right of the east/west path

The damage you can see in the above photo is far more on the east side of the stones than other sides. Again, you are not looking at normal weathering of a hard stone like granite, but some extreme situation. As we travel up and down the path, we see such damage on each of the stones.

After Karnak, we move on to another area of ancient Thebes where we will explore other strange scorch marks and seemingly cataclysmic damage. It is possible that such a plasma strike was restricted to this area, or could have been more extensive.

None of these columns would have been here at the time of the event

Luxor Temple is a large Ancient Egyptian temple complex located on the east bank of the Nile River in the city today known as Luxor (ancient Thebes). The dynastic aspects were constructed approximately 1400 BC, and it is known in the Egyptian language as ipet resyt, or 'the southern sanctuary.'

The probable path of destruction

In Luxor there are several great temples on the east and west banks. Four of the major mortuary temples visited by early travelers and tourists include the Temple of Seti I at Gurnah, the Temple of Hatshepsut at Deir el Bahri, and the Temple of Ramesses II.

Some of the massive sandstone columns at Luxor

It is from the name Luxor that we get the English word 'luxury.' Luxor Temple was built with sandstone from the Gebel el-Silsila area, which is located in south-western Egypt, less than 100 kilometers.

from Luxor. This sandstone from the Gebel el-Silsila region is referred to as Nubian Sandstone and was used for the construction for monuments in Upper Egypt, as well as past and current restoration works.

More of the sandstone columns of Luxor

The sandstone constructions both at Luxor and Karnak, including the columns seen above, were made in sections - sizes that were small enough for a group of workers to handle. However, other massive artifacts at Luxor, like we saw at Karnak, were made from granite, and likely were found there by the dynastic people when they first arrived, as they could not have made them.

Massive obelisk at the entrance to the Luxor Temple

The massive obelisk seen in the photo above was made from one piece of granite, not sections. It is estimated to weigh about 230 tons. Its twin, though not the exact same, was 'gifted' to France in the 19th century. They are both attributed to

Ramses II, but the only contribution that he made was having his name put on them, as his workers could not have cut, shaped, or moved them.

Two of the massive standing granite statues attributed to Ramses II

It was thanks to the careful observations of Yousef Awyan that further evidence that Ramses II did not have the granite works made for him were revealed. It appears that Ramses' workers, or perhaps priests, found these massive objects at Luxor, either standing as they do now, or lying on or even under the ground. Ramses was asked if he wanted his cartouche carved into them, and of course he said "Yes – many times." As you can see in the photo below, the cartouche cuts into a sword.

Note how the cartouche cuts into the sword

In the next photo, we see one of two huge one-piece granite statues at the entrance to Luxor. Each weighs several hundred tons and show damage not typical of simple defacement, but more likely catastrophic effect.

It was master machinist Chris Dunn, author of Lost Technologies of Ancient Egypt, that noticed an exceptional degree in precision in some of the works at Karnak and other locations in Egypt.

One of two massive one-pieces, minus the base granite statues

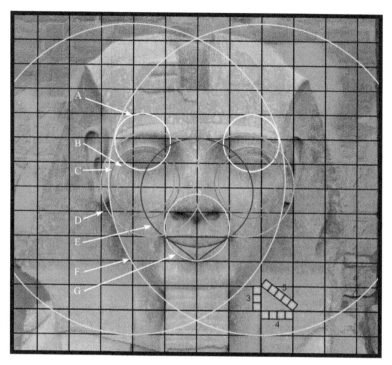
From Chris Dunn's exceptional book

He found perfect bilateral symmetry in some of the 'Ramses' faces, as well as machining marks. Such symmetry is not what a human face has, and the work would be practically impossible for modern sculptors to achieve, using machine tools, and clearly not possible by the dynastic people either.

One of the headdresses that shows astonishing precision

He was also attracted to various broken headdresses on display at Luxor and found them to be profoundly precise. Though you are not supposed to touch these, as they are roped off, one cannot help but touch the smooth - and in some cases glossy - surfaces. In the one above and others, no flaws or mistakes made by the presumed sculptor could be felt - the curves would be what you find on a modern luxury sports car.

One of the Colossi of Memnon

Not far from Luxor, one finds the Colossi of Memnon. The twin statues supposedly depict Amenhotep III (14th century BC) in a seated position, his hands resting on his knees and his gaze facing eastwards (actually ESE in modern bearings) towards the river. Two shorter figures are carved into the front throne alongside his legs. These are his wife Tiy, and mother

Mutemwiya. The side panels depict the Nile God Hapy. The statues are made from blocks of quartzite sandstone, which was quarried at el-Gabal el-Ahmar (near modern-day Cairo) and transported 675 kilometers (420 miles) overland to Thebes. (They are too heavy to have been transported upstream on the Nile.) The blocks used by later Roman engineers to reconstruct the northern colossus may have come from Edfu (north of Aswan). Including the stone platforms on which they stand, themselves about 4 meters (13 feet), the colossi reach a towering 18 meters (60 feet) in height and weight of an estimated 720 tons each.

As we have already seen, quartzite could not have been shaped in the 14th century BC, and how were such literal colossi moved, being 720 tons each? The fact that they face ESE, or about 23 degrees off to true east, like the path at Karnak, is interesting. It could very well be that these, again, are extremely ancient sculptures. If they predate 12,000 years ago, and faced true east at that time, then they could be another example of the destructive force of a plasma ejection from the sun, as we saw at Karnak.

The author with the left Colossi

The damage to the front of both colossi is extreme, and hardly likely to be the result of the elements or willful destruction. In 27 BC, a large earthquake reportedly shattered the northern colossus, collapsing it from the waist up and cracking the lower half. Following its rupture, the remaining lower half of this statue was then reputed to 'sing' on various occasions, always within an hour or two of sunrise, usually right at dawn. The

Note the lesser damage to the side and almost none at the back

sound was most often reported in February or March, but this is probably more a reflection of the tourist season rather than any actual pattern. (46) The problem with this theory is that most of the damage is on the front of both massive sculptures, with much less on the sides and virtually nothing on the back. If it was a plasma strike coming from the east, the frontal damage would have been the worst.

Path of destruction

As well, if we follow the possible course of the plasma blast towards the west and behind the Colossi, there are a series of heavily damaged sculptures, some having been recently excavated from underground and others still being searched for.

When we travel farther west, we come to the Ramesseum, where more strange damage can be found. Jean-François Champollion, who visited the ruins of the site in 1829, and first identified the hieroglyphs making up Ramesses's names and titles on the walls, coined the name, or at least its French form 'Rhamesséion.'

Possible scorching of a granite or syenite block

It was originally called the House of Millions of Years of Usermaatra-setepenra that unites with Thebes-the-city in the Domain of Amon. Most of the Ramesseum was constructed of sandstone, but there are granite elements as well.

The previous block is but one of many that show the same kind of scorching and in some cases crystal expansion that we viewed at Karnak. Our geologist Suzan Moore observed many of these stones with me, and thought that we could be witnessing the same phenomenon. She did warn that more investigation needed to be done, which is what will happen on our next trip with the Khemit School in March of 2017. The original constructions at the Ramesseum appear to be two black granite syenite gates on either side of the central sandstone complex, as well as a statue, perhaps the largest ever created. It is of granite. There are and were other granite sculptures as well, parts of which still remain, and the head and torso of one which is in the British Museum.

All of these presumed original works showed signs of weathering far more extensive than one work expect for simply having been exposed to the sun, wind, and rain since dynastic times. All showed cracks and scorch marks, as if they had been struck by a sudden blast of heat.

Badly disintegrated left gate that has been reconstructed

Portion of the remains of the right gate with seeming internal crystal expansion

Kurt posing to show a sense of scale of the torso and head of the huge sculpture

As you can see, only fragments of the base and torso remain of the syenite statue of the enthroned pharaoh, 62 feet (19 meters) high and weighing more than 1000 tons. This was alleged to have been transported 170 miles over land, but how? There are other parts of the statue still on site, but no effort to reconstruct it has been made, and likely never will. The back side has the same scorching as seen on the gates and other stones suggesting, if it was struck by plasma and then broke into pieces, that it was originally facing west, into the area of the Valley of the Kings. It would appear that both the Valley of the Kings and Valley of the Queens, though used by the dynastic people for burial, were in existence long before them. The evidence that suggests this are the tunnel systems that pervade this area.

View of the other side of the giant broken statue - notice scorching on the under side

Alcove cut into the bedrock near the Valley of the Queens

Descending into an underground tunnel system near the Valley of the Queens

System of tunnels and chambers that could connect with the Valley of the Kings

For our last explorations in Egypt, we go further south to Aswan, home to the famous granite quarry, and Elephantine Island.

First of all, there is a site near Aswan called the Tombs of the Nobles. The city of Aswan during the ancient times was not the city we know of today, as people at that period used to center around Elephantine Island, where the rulers and kings of Nubia resided. This was why the tombs of the kings and the royal family of Nubia were located near the island of Elephantine in what is called today the Tombs of the Nobles of Aswan.

View of the Tombs of the Nobles from Aswan

What is remarkable about these 'tombs' is that they are also a series of tunnels and chambers in the bedrock, going in some cases for hundreds of feet. As the stone of the area is quartzite, they could not have been made by the dynastic Egyptians. During the possible plasma strikes that may have struck the area of Thebes and possibly here to, survivers of the plasma would have used them for shelter.

Inside one of the limestone chambers

Known to the Ancient Egyptians as Abu or Yebu, the island of Elephantine stood at the border between Egypt and Nubia. It was an excellent defensive site for a city, and its location made it a natural cargo transfer point for river trade. Elephantine was a fort that stood just before the first cataract of the Nile. During the Second Intermediate Period (1650 - 1550 BC), the fort marked the southern border of Egypt. Artifacts dating back to predynastic times have been found on Elephantine, such as the one in the photo below.

A large granite stone object being pointed out by Yousef Awyan

It is likely that this object was made here, as the island itself and surrounding area are solid granite, but what it's function could have been is unknown. Of course, we presume that it is predynastic, as the precision shown in the surfaces could not have been done with bronze tools. Why it is lying on its side is also unkown. Was it dragged here to be broken up during dyastic times and recycled, or did some great force knock it over?

The author and others inspecting this masterpiece

View of the other side

A similar object broken up for other uses

Our final site in Egypt is the great quarry at Aswan. It covers an area of some 150 km^2 on both banks of the Nile from the Old Aswan Dam in the south to Wadi Kubbaniya in the north. Within this region are the famous Aswan granite quarries, less well-known ornamental silicified sandstone quarries, recently discovered extensive grinding stone quarries, as well as building stone quarries in Nubian sandstone.

The unfinished obelisk is the largest known ancient obelisk and is located in the northern region of the stone quarries.

Egyptologists believe that it was ordered by Hatshepsut (1508 - 1458 BC), possibly to complement the Lateran Obelisk (which was originally at Karnak, and was later brought to the Lateran Palace in Rome). It is nearly one third larger than any ancient Egyptian obelisk ever erected.

Dolerite stone balls believed to be the main tools

If finished, it would have measured around 42 meters (approximately 137 feet) and would have weighed nearly 1,200 tons. Hatshepsut lived several hundred years before the presence of iron or steel in Egypt, and thus was not responsible for this work.

Very deep test pit hole said to have been made by dolerite stone balls

Supposedly several years ago Egyptologist Mark Lehner spent five hours in the Aswan quarry with a dolerite hammer stone pounding against the granite bedrock (copper is too soft to cut granite). He was trying to prove that the ancient tools could do the job. He managed to excavate a one-foot square hole one-inch deep for his efforts. And yet, the video that is played in a hall at the Aswan quarry site still portrays that the hewing of the stone for the unfinished and all other obelisks was done this way.

As the previous photo shows, and the experiments of Dr. Lehner reveal, there is no way that simple stone pounders could possibly have been the main tool to quarry and shape the granite obelisks. The hole seen on the previous page is several feet deep, and there is no space for the worker to descend into the hole longer than his arm length. There are several of these 'test pits' that would appear to have been made to see how sound and consistent the rock was. Two distinct patterns appear in the quarry as evidence of ancient people working the stone.

Dynastic Egyptian and later technique for splitting granite

The above photo shows how granite was split in dynastic Egypt, as well as being the likely technique of later Greek and Roman people. The grooves were made possibly with very harndened bronze tools that had to be sharpened several times, and then wooden wedges were inserted. After achieving the tightest fit possible, water, likely hot, was poured on the wedges causing them to expand over time and hopefully causing the stone to split along the desired line.

In contrast, the tool marks in the area of the giant unfinished obelisk as well as another smaller one are completely different.

'Scoop marks' present on horizontal and vertical surfaces in the quarry

There are thousands of marks in the surfaces of the quarry that are depressions of varying sizes and widths. Best desribed as scoop marks, they are on the horizontal and vertical surfaces, as well as undercuts at both of the unfinished obelisks. The idea that these were made by dolerite stone balls becomes laughable, especially when you see the depth of the great unfinished obelisk's trenches, and the undercuts on the smaller obelisk.

The sheer 1200 ton size of the great unfinished obelisk

Master machinist Chris Dunn suggested that this was done by some kind of machine that he describes as a massive belt sander, attached to some kind of excavator. Another theory that the author supports is that the ancient builders had a device that emitted a vibrational wave tuned to disrupt either the quartz or feldspar in the stone, much like how an opera singer can shatter a crystal wine glass. Such a technology could literally turn hard stone into sand.

The left channel of the great unfinished obelisk

Why the two obelisks were never finished is unknown. It appears that the workers simply stopped and never came back. The same possibly was the case at the Serapeum at Saqqara, where most of the 100 ton boxes were never finished. If these are pre-dynastic works as expected, then the great cataclysm of 12,000 years ago could have been the culprit - massive earthquakes and possibly solar blasts devastating all life in these and other areas. There is a massive horizontal crack in the bottom of the right channel of the great obelisk that also may have been the reason.

Crack in the bedrock in the right channel of the great obelisk

If this and the other obelisks were originally used as resonance devices, then the encountering of such a crack above the intended final depth of the obelisk would render it useless as a vibratory instrument. A similar crack is found on the smaller obelisk that was never completed.

Photo showing the narrowness of the smaller obelisk's channels

On another note, the left and right channels of the great obelisk would have been too narrow for workers to be shaping them with dolerite stone pounders. A lot of force would be required to remove any material at all, and having a foot or two of clearance would result in very little if any stone removal.

Engineer Tony inspecting a depression where a huge stone box may have been removed

This quarry was not just for the making and planned removal of the two obelisks. In the above photo, mining engineer Tony from Australia found the depression above where perhaps a huge stone box had been extracted, as well as another from where a large statue had perhaps been successfully removed. Unfortunately, the remaining quarry area that can be explored has been reduced over the years by housing and commercial developments, which could be sitting right on top of a lot more ancient evidence.

Shaped 'staircase' leading down to the smaller obelisk

One must also take into account that the creation of the Aswan dams caused flooding of quite vast parts of the Aswan area. It could very well be that granitic stone of finer consistency, such as diorites, may have been harvested in pre-dynastic and even dynastic times in quarry areas that are now underwater. It is hoped that on further trips with the Khemit School Suzan Moore, and other geologists, may be able to find more evidence that we can use in future books, articles, and videos.

Area where a massive obelisk may have been removed

Deep and even scoop marks on the smaller obelisk

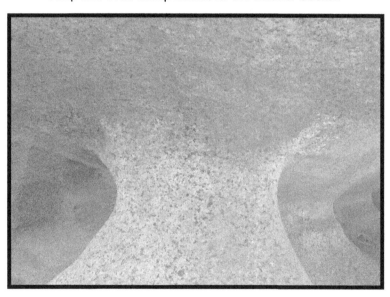
Undercuts showing that the workers were close to releasing the smaller obelisk

Members of the Khemit School trying to release the smaller obelisk to no avail

4. Lebanon

As regards Lebanon, we will focus on one site, that of Baalbek. Baalbek, properly Baʿalbek and also known as Balbec, Baalbec, or Baalbeck, is a town in the Anti-Lebanon foothills east of the Litani River in Lebanon's Beqaa Valley, about 85 kilometers (53 miles) northeast of Beirut and about 75 kilometers (47 miles) north of Damascus. It presently has a population of approximately 82,608, mostly Shia Muslims, followed by Sunni Muslims and a minority of Christians. In Greek and Roman antiquity, it was known as Heliopolis. It still possesses some of the best preserved Roman ruins in Lebanon, including one of the largest temples of the empire. The gods that were worshipped there (Jupiter, Venus, and Bacchus) were equivalents of the Canaanite deities Hadad, Atargatis, and another young male fertility god. Local influences are seen in the planning and layout of the temples, which vary from the classic Roman design.

Roman-style Temple of Hercules at Baalbek

The hilltop of Tell Baalbek, part of a valley to the east of the northern Beqaa Valley, shows signs of almost continual habitation over the last 8000 - 9000 years. It is well watered both from a stream running from the Rās-el-ʿAin spring SE of the citadel and, during the spring, from numerous rills formed by melt water from the Anti-Lebanons. Macrobius later credited the site's foundation to a colony of Egyptian or Assyrian priests. (47)

The settlement's religious, commercial, and strategic importance was minor enough, however, that it was never mentioned in any known Assyrian or Egyptian record, unless under another name. Its enviable position in a fertile valley, major watershed, and proximity to the route from Tyre to Palmyra should have made it a wealthy and splendid site from an early age.

During the Canaanite period, the local temples were largely devoted to the Heliopolitan Triad - the male God Baʿal, his consort Ashtart, and their son Adon. (48) Following Alexander the Great's conquest of Persia in the 330s BC, Baalbek (under its Hellenic name Heliopolis) formed part of the Diadochi kingdoms of Egypt and Syria, later annexed by the Romans during their eastern wars. The settlers of the Roman colony Colonia Julia Augusta Felix Heliopolitana may have arrived as early as the time of Caesar but were more probably the veterans of the 5th and 8th Legions under Augustus, during which time it hosted a Roman garrison. From 15 BC to 193 AD it formed part of the territory of Berytus, and it is believed that during this time most of the major construction was done. (49)

The sheer size of some of the limestone foundation blocks are the largest ever quarried on the planet, conservatively estimated at 800 to 1200 tons, and the common belief that the Romans chose to do this work on such a massive scale to 'impress the locals' is absolutely ludicrous. Nowhere else in the Roman world is there any evidence of the quarrying of blocks of this size, so we can clearly presume that they were there when the Romans first appeared, and were used as foundational material.

A group of three horizontally lying giant stones which form part of the podium of the Roman Jupiter Temple of Baalbek, Lebanon, go by the name 'trilithon.' Each one of these stones is 70 feet long, 14 feet high, 10 feet thick, and weighs around 800 tons. These three stone blocks are the largest building blocks ever used by any human beings anywhere in the world. (50) The supporting stone layer beneath features a number of stones that are still in the order of 350 tons and 35 feet wide.

Some of the massive Baalbek blocks

No one knows how these blocks were moved, cut, placed, and fit perfectly together. Many like to say aliens were involved since they are so heavy, and seemingly impossible for ancient humans to move. But there is another theory developed by Jean-Pierre Adam in his article "A propos du trilithon de Baalbek. Le transport et la mise en oeuvre des megaliths" (About Trilithon Baalbek:

Transmission and Implementation of Megaliths.) He came up with two basic yet complex tools or mechanisms to be able to transport the stones. The first, very simply put, consisted of a very wide metal wheel, surrounded by a wooden platform. The second method of transport also simply put involved two wooden wheels, 12 feet apart, with a large iron platform in between. Horses would have drawn both of these. He then went on to describe a large pulley system operated by many men, consisting of large wooden wheels, with poles attached, that the men would push. This force would turn the wheel, pulling the cable made of hemp, which in turn would move the large blocks of trilithon.

Absolutely ridiculous.

You can see from photos of the trilithon that they are far larger than required for foundation blocks, and in fact do not match the rest of the foundation.

Trilithon blocks with smaller ones below

Also, they are the largest stones in the entire construction, and are larger than the stones below. The only other blocks of similar size in the area are two, which still lie in the quarry about a mile away from the trilithon area. One, estimated at about 1000 tons, was released from the bedrock and seemingly dropped down to the ground below, being partially buried at one end. There is also clear evidence that someone, perhaps the Romans, attempted to break it into smaller pieces, with limited success.

1000 ton block in the Baalbek quarry

The other is at the other side of the quarry - the famous 'stone of the pregnant woman.' Estimates of its size are between 1200 and 1500 tons, and it is still attached to the bedrock. There is a crack where the shaped stone area meets the bedrock base. Whether this was part of the quarrying technique or a natural fault is unknown. And next to it, recent excavations have exposed what is either an even more massive stone block under the Earth, or the bedrock itself. The author, upon personal inspection, believes that in this case, the stone was of a low quality, and thus attempts to shape it into a large block were abandoned.

Stephen Mehler and the author in the quarry with the 'pregnant woman'

There are also up to 300 rose granite pillars in various states of damage at the main Baalbek site, and massive limestone columns also largely damaged from hundreds if not thousands of years of conflict in the area. The limestone columns could clearly have come from the local quarry, and though some are up to 40 feet tall, they were made in sections. This would mean that the Roman stone masons, who presumably could have had iron and even steel tools, could have shaped the sections in the quarry and then rolled them the one mile to the Baalbek temple site.

Limestone pillar sections

As regards the rose granite columns, however, that is a completely different story. Granite of this type is not found naturally anywhere in the Baalbek area, or indeed the country of Lebanon. Our geologist expert Suzan Moore from Canada, who joined us with the Khemit School of

Giza Egypt on our 2015 expedition to Lebanon, believed that this granite may in fact have come from Aswan, in the very south of Egypt. This is a distance of about 700 miles (1100 kilometers) and the Romans would unlikely have transported 300 of these granite columns such a distance. Also, the finish of the columns is astonishing. Even to this day, and after all of the damage to them, we saw and felt surfaces that could only have been done on a giant lathe.

Rose granite column section of near perfection

With estimates that some of these columns could have originally been 20 feet tall, and were of one solid piece, such a machine could not have existed during Roman, Greek, or dynastic Egyptian times.

This leads one to speculate that they could have been made by an even older civilization, as in Peru, Bolivia, and Egypt, that had technology beyond that of the 20th century.

As we saw in the cases of the other geographic areas we have covered, the presumed builders, such as the Inca and dynastic Egyptians, though talented people, could not have worked in hard stone well. We have seen that this is also the case in Lebanon, where the Romans, though quite adept at working marble and limestone, could not shape granite to such a fine tolerance as seen in the columns.

One candidate for who could have been responsible for making the granite columns and cutting and moving the giant limestone blocks, if they came from the Aswan quarry in Egypt, is the mysterious Khemitians, discussed earlier. The level of technical quality and capability of megalithic works matches, or in some cases surpasses, that of the pre-dynastic monumental works in Egypt. Some would contend that it was in fact the Nephilim from the Bible. The Nephilim were a race that came to dominate the antediluvian (pre-flood) world, and are referred to in the Bible as the heroes of old, men of renown. They were reportedly the children born to the 'Sons of God' by the 'daughters of men,' and are described as giants. It is also most important to note that they are mentioned almost simultaneous to God's statement that He would destroy the Earth by flood, and it seems from this association that their effect upon mankind was one of the primary justifications that brought the destruction. (51)

As the Bible, or at least the Old Testament, is not accurate as regards timeframes for the flood, and we are to guess that it could be describing 'the flood' as the great catastrophe of about 12,000 years ago, then this would make the mysterious works of Baalbek contemporary with the sites in Peru, Bolivia, and Egypt that we have explored.

'Nephilim' is rendered fallen, or possibly feller, a tyrant or bully. Several English translations such as the King James Version, rendered the word 'giants.' In the Greek Septuagint the word, 'Nephilim,' was also translated as 'gigantes' (gigantic). This translation is undoubtedly used because the Nephilim later became known as giants to the ancient Hebrews, as illustrated by the manner they were referenced when the Israelite spies were sent into Canaan (Numbers 13:33). (52) It is unclear what the Sons of God were, but they are distinguished from the daughters of men. The most obvious interpretation is that the Nephilim were a hybrid race between two distinct beings. There are at least three schools of thought regarding the Sons of God.

The older view, held nearly unanimously by ancient writers prior to Augustine of Hippo, is that the Nephilim were a hybrid race between certain fallen angels, called the Benei Ha'Elohim ('Sons of God') or The Watchers in extra-Biblical traditions, and human women. While there has always been a minority of churchmen who followed this view, it has been promoted recently by popular writers such as Stephen Quayle. (53) The more recent view that has been the majority position in the church since St. Augustine in the 4th century is that the Sons of God refers to the god-fearing line of Seth, and the daughters of men refers to the daughters of the unbelieving line of Cain. Variations on this theme include the idea, proposed by Meredith Kline, that the Sons of God were kings or priests who took any woman they chose to be their wife.

Still others hold that the Sons of God were other created men. It is argued that the Bible does not describe every person that was created, but only key individuals or situations are included within the text. Those holding to this position call into question the origin of Cain's wife or those whom he feared would kill him (Genesis 4:14-17). However, this view falls into conflict with Genesis, which states that Eve is the mother of all the living.

Whether the Nephilim ever existed at all is a matter of conjecture. As no skeletal remains have presumably been on public display at least in recent times, there does not seem to be hard evidence that they were in fact living beings. Thus, the creators of Baalbek, or at least those that laid the foundation work that was done prior to Roman times, remain a mystery, as do those in Peru, Bolivia, Egypt, and other places where we find evidence of lost ancient high technology.

The commonalities that we find at all of the above locations include: 1) works in stone beyond the scale and technical prowess of the historically presumed builders, such as the Inca; 2) signs of construction interruptions and/or cataclysmic damage;

and 3) oral traditions speaking of much earlier civilizations with advanced technological capabilities.

There are other places of interest pertaining to this subject, including Turkey, Greece, the western part of Italy, Angkor Wat, etc., but as the author has not been there in person, they will perhaps be the subject of another book. Also, widespread claims of massive pyramids and other structures having been built in such locations as Siberia, Antarctica and others have seemingly not been backed up by actual scientific evidence. Just because a hill or mountain appears to be pyramidal in shape does not make it an actual pyramid, and any cultural context is sadly lacking in most cases.

6. Closing Thoughts

It is hoped that from what you have read in this book, and the evidence presented, supports the argument that the history of humanity must be rewritten to include a period of history either ignored - or neglected - by most academics. By incorporating the scientific method, and the knowledge of geologists, engineers, machinists, chemists, physicists, and other experts, as well as accepting that the oral traditions of indigenous people is their history, then our view of the past will radically change.

Rather than humanity living as relatively sparsely spread-out hunter gathers for the vast majority of its existence, we can now see that there was a period of time, or perhaps periods, when the ancestors of this planet were in fact more highly developed than we are - technologically and perhaps spiritually as well. Thankfully, our own modern technologies, including the internet and the fact that most people carry the ability to record high definition photos and video on their phones, facilitate more people having the capability of collecting and sharing interesting data than ever before.

The time is now gone for exploration and intellectual inquiry to be the sole domain of academics, who rarely share their results with the public. Anyone can be a researcher, and hopefully will broadcast whatever they learn to the global population so that we, as a human family, can see how profound our collective history is, and how amazing our future, together, can be.

Brien Foerster
Paracas Peru, 2016

6. Bibliography

1. Leeming, David (2004). "Flood: The Oxford Companion to World Mythology." Oxfordreference.com.
2. Turney, C.S.M.; Brown, H. (2007). "Catastrophic early Holocene sea level rise, human migration and the Neolithic transition in Europe." Quaternary Science Reviews 26 (17–18): 2036–2041.
3. McGowan, Christopher; The Dragon Hunters, Cambridge, MA: Perseus Publishing, 2001, ISBN 0-7382-0282-7.
4. Rudwick, Martin J. S.; The Meaning of Fossils, Chicago, IL: The University of Chicago Press, 1972, ISBN 0-226-73103-0 pp. 133 to 134.
5. Rudwick, Martin J. S.; The Meaning of Fossils, Chicago, IL: The University of Chicago Press, 1972, ISBN 0-226-73103-0 pp. 131.
6. Rudwick, Martin J. S.; The Meaning of Fossils, Chicago, IL: The University of

Chicago Press, 1972, ISBN 0-226-73103-0 pp. 133 to 135.
7. Rudwick, Martin J. S.; The Meaning of Fossils, Chicago, IL: The University of Chicago Press, 1972, ISBN 0-226-73103-0 pp. 135.
8. Rudwick, Martin J. S.; *The Meaning of Fossils*, Chicago, IL: The University of Chicago Press, 1972, ISBN 0-226-73103-0 pp. 136 to 138.
9. Rudwick, Martin J. S.; The Meaning of Fossils, Chicago, IL: The University of Chicago Press, 1972, ISBN 0-226-73103-0 pp. 174 to 175.
10. Rudwick, Martin J. S.; *The Meaning of Fossils*, Chicago, IL: The University of Chicago Press, 1972, ISBN 0-226-73103-0 pp. 174 to 179.
11. Krystek, Lee. "Venus in the Corner Pocket: The Controversial Theories of Immanuel Velikovsky". Museum of Unnatural Mystery.
12. http://www.bookreview.com/spindb.query.listreview2.booknew.7892.

13. https://rajarasablog.wordpress.com/tag/allan-delairs-cataclysm-11500-years-ago/.
14. http://www.knowth.com/sacred-geography-1.htm.
15. http://community.humanityhealing.net/profiles/blogs/our-planetary-journey-from-catastrophobia-to-spiritual-awakenin-2.
16. Burbridge, G. R. et al. "Evidence for the occurrence of violent events in the nuclei of galaxies." Reviews of Modern Physics 35 (1963): 947.
17. Oort, J. H. "The Galactic Center." Annual Reviews of Astronomy & Astrophysics 15 (1977): 295.
18. Lo, K. Y., and Claussen, M. J. "High-resolution observations of ionized gas in central 3 paresecs of the Galaxy: possible evidence for infall." Nature 306 (1983): 647.
19. http://beforeitsnews.com/books/2014/10/earth-under-fire-by-dr-paul-laviolette-2483512.html.

20. Anthony D. Barnosky, Paul L. Koch, Robert S. Feranec, Scott L. Wing, Alan B. Shabel (2004). "Assessing the Causes of Late Pleistocene Extinctions on the Continents." *Science* 306 (5693): 70–75.
21. Scott, E. (2010). "Extinctions, scenarios, and assumptions: Changes in latest Pleistocene large herbivore abundance and distribution in western North America." *Quat. Int.* 217: 225.
22. MacFee, R.D.E. & Marx, P.A. (1997). "Humans, hyper disease and first-contact extinctions." In Goodman, S., Patterson, B.D. *Natural Change and Human Impact in Madagascar.* Washington D.C.: Smithsonian Press. pp. 169–217.
23. MacFee, R.D.E. & Marx, P.A. (1997). "Humans, hyper disease and first-contact extinctions." In Goodman, S., Patterson, B.D. *Natural Change and Human Impact in Madagascar.* Washington D.C.: Smithsonian Press. pp. 169–217.

24. http://www.aip.org/history/climate/rapid.htm#S7A
25. http://www.aip.org/history/climate/rapid.htm#S7A
26. Firestone RB, West A, Kennett JP, *et al.* (October 2007). "Evidence for an extraterrestrial impact 12,900 years ago that contributed to the megafauna extinctions and the Younger Dryas cooling." *Proc. Natl. Acad. Sci. U.S.A.* 104 (41): 16016–21.
27. Kennett DJ, Kennett JP, West A, *et al.* (January 2009). "Nano diamonds in the Younger Dryas boundary sediment layer." *Science* 323 (5910): 94.
28. Wittke, James H. (2013-05-20). "Evidence for deposition of 10 million tonnes of impact spherules across four continents 12,800 y ago."
29. Hubbe A., Hubbe M., Neves W (2007). "Early Holocene survival of megafauna in South America." *Journal of Biogeography* 34 (9): 1642–1646.
30. http://www.angelfire.com/empire2/unkemptgoose/Civilization.html.

31. http://www.angelfire.com/empire2/unkemptgoose/Civilization.html.
32. Fagan, Brian M. *The Seventy Great Mysteries of the Ancient World: Unlocking the Secrets of Past Civilizations*. New York: Thames & Hudson, 2001.
33. (a) Kolata, Alan L. (December 11, 1993). *The Tiwanaku: Portrait of an Andean Civilization*. Wiley-Blackwell. ISBN 978-1-55786-183-2.
    (b) Kolata, Alan L. *Valley of the Spirits: A Journey into the Lost Realm of the Aymara,* Hoboken, New Jersey: John Wiley and Sons, 1996.
    (c) Wright, Kenneth R.; McEwan, Gordon Francis; Wright, Ruth M. (2006). *Tipon: Water Engineering Masterpiece of the Inca Empire.*
    (d) Foerster, Brien 2011. *A Brief History Of The Incas: From Rise, Through Reign To Ruin.* Hidden Inca Press.
34. Shaw, Ian (2003). *The Oxford History of Ancient Egypt*. Oxford, England: Oxford University Press. ISBN 0-19-280458-8.

35. http://www.reshafim.org.il/ad/egypt/trades/metals.htm.
36. http://www.reshafim.org.il/ad/egypt/trades/metals.htm.
37. Coins of Septimius Severus bear the legend col·hel·i·o·m·h: Colonia Heliopolis Iovi Optimo Maximo Helipolitano.
38. http://www.touregypt.net/wadihammamat.htm.
39. http://www.messagetoeagle.com/mysterious-piece-of-sophisticated-technology-could-rewrite-history-scientists-are-not-sure-what-they-are-dealing-with/.
40. Dee, M.; Wengrow, D.; Shortland, A.; Stevenson, A.; Brock, F.; Girdland Flink, L.; Bronk Ramsey, C. (4 September 2013). "An absolute chronology for early Egypt using radiocarbon dating and Bayesian statistical modeling." Proceedings of the Royal Society A: Mathematical, Physical and Engineering Sciences 469 (2159).
41. http://www.ancientmysteries.com/testpart/valleytemp/ValleyTemple/valleytemple.html.

42. http://www.robertschoch.com/sphinxcontent.html.
43. Romer, John (2007). The Great Pyramid: Ancient Egypt Revisited. Cambridge University Press, Cambridge.
44. Mathieson, I., Bettles, E., Dittmer, J., & Reader, C. (1999). The National Museums of Scotland Saqqara survey project, Earth sciences 1990-1998. Journal of Egyptian archaeology, 85, 21-43.
45. https://books.google.com.pe/books?vid=0wPPdxNrZi0_mMbp&id=xznqw5EkOCMC&redir_esc=y&hl=en.
46. https://en.wikipedia.org/wiki/Colossi_of_Memnon.
47. "Helio'polis Syriae," A Dictionary of Greek and Roman Geography, Vol. I, London: John Murray, 1878, pp. 1036–1038.
48. Jidejian, Nina (1975), Baalbek: Heliopolis: "City of the Sun," Beirut: Dar el-Machreq Publishers, ISBN 978-2-7214-5884-1.

49. "Baalbec," Encyclopædia Britannica, 9th ed., Vol. III, New York: Charles Scribner's Sons, 1878, pp. 176–178.
50. "Baalbek Trilithon." About.com Religion & Spirituality.
51. http://nwcreation.net/nephilim.html.
52. Morris, Henry M., The Genesis Record. Grand Rapids MI: Baker Books, 1976. p.172.
53. Stephen Quayle, Genesis 6 Giants, http://stevequayle.com.

Made in the USA
Middletown, DE
08 February 2019